Biodegradability Prediction

NATO ASI Series

Advanced Science Institutes Series

A Series presenting the results of activities sponsored by the NATO Science Committee, which aims at the dissemination of advanced scientific and technological knowledge, with a view to strengthening links between scientific communities.

The Series is published by an international board of publishers in conjunction with the NATO Scientific Affairs Division

A	**Life Sciences**	Plenum Publishing Corporation
B	**Physics**	London and New York
C	**Mathematical and Physical Sciences**	Kluwer Academic Publishers
D	**Behavioural and Social Sciences**	Dordrecht, Boston and London
E	**Applied Sciences**	
F	**Computer and Systems Sciences**	Springer-Verlag
G	**Ecological Sciences**	Berlin, Heidelberg, New York, London,
H	**Cell Biology**	Paris and Tokyo
I	**Global Environmental Change**	

PARTNERSHIP SUB-SERIES

1.	**Disarmament Technologies**	Kluwer Academic Publishers
2.	**Environment**	Springer-Verlag / Kluwer Academic Publishers
3.	**High Technology**	Kluwer Academic Publishers
4.	**Science and Technology Policy**	Kluwer Academic Publishers
5.	**Computer Networking**	Kluwer Academic Publishers

The Partnership Sub-Series incorporates activities undertaken in collaboration with NATO's Cooperation Partners, the countries of the CIS and Central and Eastern Europe, in Priority Areas of concern to those countries.

NATO-PCO-DATA BASE

The electronic index to the NATO ASI Series provides full bibliographical references (with keywords and/or abstracts) to more than 50000 contributions from international scientists published in all sections of the NATO ASI Series.
Access to the NATO-PCO-DATA BASE is possible in two ways:

– via online FILE 128 (NATO-PCO-DATA BASE) hosted by ESRIN,
Via Galileo Galilei, I-00044 Frascati, Italy.

– via CD-ROM "NATO-PCO-DATA BASE" with user-friendly retrieval software in English, French and German (© WTV GmbH and DATAWARE Technologies Inc. 1989).

The CD-ROM can be ordered through any member of the Board of Publishers or through NATO-PCO, Overijse, Belgium.

Series 2: Environment – Vol. 23

Biodegradability Prediction

edited by

Willie J. G. M. Peijnenburg

National Institute of Public Health and the Environment,
Bilthoven, The Netherlands

and

Jirí Damborský

Department of Microbiology, Faculty of Science,
Masaryk University,
Brno, Czech Republic

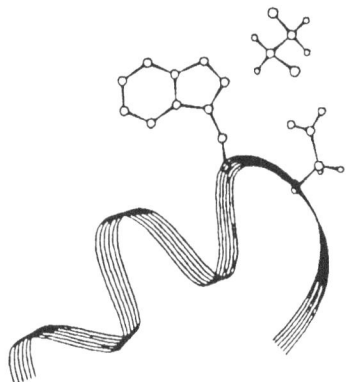

Springer-Science+Business Media, B.V.

Proceedings of the NATO Advanced Research Workshop on
QSAR Biodegradation II: QSARs for Biotransformation & Biodegradation
Luhačovice, Czech Republic
May 2–3, 1996

A C.I.P. Catalogue record for this book is available from the Library of Congress

ISBN 978-0-7923-4341-7   ISBN 978-94-011-5686-8 (eBook)
DOI 10.1007/978-94-011-5686-8

Printed on acid-free paper

TABLE OF CONTENTS

Preface   vii

Acknowledgement   ix

Introduction

Introduction, Main Conclusions and Recommendations of the Workshop
"QSAR Biodegradation II"
    W. J. G. M. Peijnenburg, J. Damborský   1

Part I. Biodegradability (foundations, testing)

Biodegradability of Xenobiotic Organic Compounds Depends on their Chemical
Structure and Efficiently Controlled, and Productive Biochemical Reaction
Mechanisms
    R.-M. Wittich   7
Biodegradability Testing of Xenobiotics
    P. Pitter, V. Sýkora   17

Part II. Biodegradability Modelling (trends, methods)

The META-CASETOX System for the Prediction of the Toxic Hazard
of Chemicals Deposited in the Environment
    G. Klopman   27
Application of Artificial Intelligence in Biodegradation Modelling
    D. Gamberger, S. Sekušak, Ž. Medven, A. Sabljić   41
Polychlorinated Dibenzo-p-Dioxins in Anaerobic Soils and Sediments. A Quest
for Dechlorination Pattern-Microbial Community Relationships
    P. Adriaens, A.L. Barkovskii, M. Lynam, J. Damborský, M. Kutý   51
A Biodegradability Evaluation and Simulation System (BESS) Based on
Knowledge of Biodegradation Pathways
    B. Punch, A. Patton, K. Wight, R. J. Larson, P.H. Masscheleyn, L. Forney   65
A Mechanistic Approach to Deriving Quantitative Structure Biodegradability
Relationships. A Case Study: Dehalogenation of Haloaliphatic Compounds
    J. Damborský, K. Manová, M. Kutý   75

Part III. Biodegradability Prediction (applications)

Quantitative Structure-Biodegradability Studies: An Investigation of the MITI
Aromatic Compound Data-Base
    J. Dearden, M. T. D. Cronin   93

Prediction of Biodegradability from Chemical Structure: Use of MITI Data,
Structural Fragments and Multivariate Analysis for the Estimation of
Ready and Not Ready Biodegradability
    H. Loonen, F. Lindgren, B. Hansen, W. Karcher   105
Development of Structure-Biodegradability Relationships (SBRs) for
Estimating Half-Lifes of Organic Contaminants in Soil Systems
    R. Govind, L. Lei, H. Tabak   115

Subject Index   139

Author Index   143

PREFACE

Large numbers of hazardous organic chemicals are emitted into the environment from anthropogenic and natural sources. Assessment of the ecological impact of the emissions related to the production and use of chemicals, requires to have available simple, and above all, reliable models that can be used to both qualitatively and quantitatively describe the fate and behaviour of these compounds in the environment. Up till now several of the transport and transformation processes that may contribute significantly to the removal of xenobiotics from distinct environmental compartments have been studied in detail and various models and process descriptions are available. Several reaction paths have been investigated intensively and an increasing number of reaction rate constants can now be estimated quite accurately using a variety of Quantitative Structure Activity Relationships (QSARs).

For most organic chemicals, biodegradation is the dominant transformation pathway, contributing significantly to the attenuation of their environmental concentrations. Presumably due to the complexity of the biodegradative pathways, at the moment it only to a limited extent is possible to extrapolate results of biodegradation tests obtained in the laboratory, to realistic environmental field conditions; methods for predicting rates of biodegradation in the field seem to be lacking completely. Thus, to understand the fate of a substance and to predict exposure concentrations for risk assessment, information on the biodegradability of a substance is critical to the assessment. Currently, however, QSARs for estimating the biodegradability of a substance are not only limited in number, but their validity is being questioned.

These observations prompted us to organise a first workshop on QSARs for Biodegradation in September 1994 (as a satellite workshop to the 6th International Workshop on QSAR in Environmental Sciences, organised in Belgirate [Italy] by dr. W. Karcher - European Chemicals Bureau, Joint Research Centre of the European Communities, Ispra [Italy]). Although one of the main conclusions was that there still is insufficient knowledge of the basic processes that take place during biodegradation, at the same time the meeting revealed that a broad range of molecular descriptors is available with high potency to be the relevant parameters for description of the structural features connected with the ability of a compound to be converted by the action of living organisms [1]. Also, highly sophisticated statistical and computational methods are available, like multivariate statistics, neural networks, genetic algorithms, to explore and find structure-activity relationships. It was revealed that it is the biodegradation data entering in the analysis and probably also the whole philosophy used for biodegradation modelling which can not be considered as being satisfactory at present. There is no doubt, that the biodegradability of a particular organic compound in the environment is not dependent only on its structure, but also on factors related to the environmental conditions the chemical resides in. The latter factors were in the past completely neglected in Quantitative Structure-Biodegradabiliy Relationship (QSBR) models. In other words, it was concluded that up till now insufficient attention has been paid to the "biological" and "ecological" part of the biodegradation process in structure-biodegradability models.

Considering the long history of the investigations into the processes of biodegradation of organic compounds by living organisms, we believe that there exists far more knowledge about these processes than has been incorporated into the structure-bio-degradability models to date. We therefore decided to organise the second workshop on QSARs for biodegradation. In co-operation with the NATO Scientific Affairs Division, this workshop was held in May 1996 in Luhacovice (Czech Republic). One of the ideas standing behind the meeting in Luhacovice was to bring together experts working on biodegradation problems and those working on the development and the analysis of Quantitative Structure-Biodegradability Relationships, and initiate a public discussion on 'bringing the knowledge on biological aspects of biodegradation into QSBR models'. As the QSAR analysis has always been considered to be a interdisciplinary area, we believe that this discussion could be successful.

In these proceedings, participants of the "NATO Advanced Research Workshop on QSAR Biodegradation II" report on the present state of art with regard to QSBRs and on the main findings of the deliberations. The contributions may be classified into three distinct categories:

(i) The main characteristics and foundations of biodegradation. This section includes on the one hand overviews of the microbial aspects of biodegradation, whereas on the other hand the methods that are available for testing the biodegradability of chemical substances are reviewed.

(ii) Trends and methods in biodegradation modelling. In this section amongst others, several computerised systems for the prediction of the biodegradability of chemicals deposited in the environment are highlighted.

(iii) The application of predictive methods for biodegradation: the application and validation of some newly developed predictive models for biodegradation of organic compounds in both the aquatic and the terrestrial compartment is highlighted in this section.

W.J.G.M. Peijnenburg,
J. Damborský,
Editors

REFERENCE

1. Peijnenburg, W.J.G.M. and Karcher, W. (eds.), *Proceedings of the Workshop "Quantitative Structure Activity Relationships for Biodegradation"*, National Institute of Public Health and Environmental Protection (RIVM), Bilthoven, The Netherlands, Report no. 719101021.

ACKNOWLEDGEMENT

On behalf of all participants to the Workshop "QSAR Biodegradation II: QSARs for Biotransformation & Biodegradation", we gratefully acknowledge all support obtained from the NATO Scientific Affairs Division. Special thanks are due to dr. L. Veiga da Cunha, Director Priority Area on Environmental Security, for his active support in the organisational aspects of the meeting. Also we wish to acknowledge the organisers of TOCOEN '96, and especially prof. I. Holoubek, for kindly providing all logistic and technical facilities during the workshop.

Part of the preparation of the workshop report was carried out under the framework of the project "Fate and Activity Modelling of Environmental Pollutants Using Structure Activity Relationships", financially supported by the Environmental Technologies RTD Programme (DG XII/D-I) of the Commission of the European Communities under contract number ENV4-CT96-0221. Additional funding was obtained in the framework of the EU Programme on Science and Technology Co-operation with Central and Eastern European countries under the supplementary agreement number CIPD-CT93-0042. Financial support from the European Union is gratefully acknowledged.

W.J.G.M. Peijnenburg
J. Damborský

INTRODUCTION, MAIN CONCLUSIONS AND RECOMMENDATIONS OF THE WORKSHOP "QSAR BIODEGRADATION II"

W.J.G.M. PEIJNENBURG
National Institute of Public Health and the Environment,
Laboratory for Ecotoxicology, P.O. Box 1, 3720 BA Bilthoven,
THE NETHERLANDS

and

J. DAMBORSKÝ
Faculty of Science, Masaryk University
Kotlárská 2, 611 37 Brno, CZECH REPUBLIC

1. Introduction

For most organic chemicals, biodegradation is the dominant transformation pathway, contributing significantly to the attenuation of their environmental concentrations. When biodegradation is complete and mineralisation to carbon dioxide or methane occurs, organic carbon and other elements from the substance, like for instance nitrogen and sulphur, are released and reassimilated into natural elemental cycles. To understand the fate of a chemical and to assess its ecological impact, information on the biodegradability is critical. Basically two types of biodegradation data are required. The first type of information is on whether the substance is completely biodegradable or if persistent metabolites are formed. In the case of formation of persistent metabolites an additional impact assessment will need to be conducted for them and the long-term potential for impacts due to accumulation evaluated. The second type of data is on the rate of biodegradation in relevant environmental compartments. This information is used to predict the concentrations to which organisms will be exposed in the environment. The rate data can also be used to assess the potential for accumulation.

The question whether a chemical is completely biodegradable is usually answered by looking at the results of a standard biodegradation test. There are several ways to estimate the rate of biodegradation in the environmental compartment in which the chemical will reside. The most straight forward method is to measure the rate in the environmental media of interest. However, these type of biodegradation tests are complicated and thus data are limited. An alternative approach is to measure the rate in a screening test or to assume some rate for substances that "pass" the test and apply a scaling factor to convert that rate to one that is applicable to the environmental compartment of interest. In case of substances that are to be considered readily biode-

1

W. J. G. M. Peijnenburg and J. Damborský (eds.), Biodegradability Prediction, 1–5.
© 1996 *Kluwer Academic Publishers.*

gradable, these scaling factors often represent the relative magnitudes of total biomass in the environmental compartment compared to that in the test.

For new chemicals, data on biodegradation is often obtained as part of the database for the registration of a new chemical [1], but for the thousands of substances currently in use biodegradation data is often unavailable or of questionable quality [2]. For these substances a method is needed to estimate the biodegradability under realistic environmental conditions from information on its structure. The types of challenge that arise in developing predictive methods for biodegradation relate to the endpoint desired. Thus, if the goal of the method is to predict the biodegradation rate in the relevant environmental compartments, the challenge is both with getting truly representative data and with the relevance of the available rate data to the real world.

Computer models based on systematic rationalisation of experimental data of a large and diverse group of chemicals can help address the problem. Several such models have been developed, providing accurate reproduction of experimental biodegradation data. The models are often based on statistical evaluation of structural features associated with biodegradation. New models, based on mechanistic consi-derations start to appear as well. At the moment it is only to a limited extent that it is possible to extrapolate results of biodegradation tests obtained in the laboratory to realistic environmental field conditions. Methods for predicting rates of biodegradation seem to be lacking completely. This is presumably due to the complexity of the biodegradation process.

Another major drawback that is encountered, is the observation that often the environmental conditions (including the presence of microbes capable of degrading a specific chemical) can be as important as structure in determining the degradability of a chemical. As an alternative, and as explained above, the question of whether a substance is biodegradable is usually answered by means of standardised laboratory biodegradation testing (e.g. according to the OECD test methods). It should be noted that the laboratory conditions do not always reflect the environmental reality and therefore the question of extrapolation of both test results as well as model predictions based on laboratory tests to the real environment remains necessary. Although by now more detailed kinetic information is becoming available, the endpoint of a standardised test often is: yes/no biodegradable after a certain testing period. Despite the fact that up until now various predictive models for biodegradation have been developed, it was shown by means of an extensive validation study that only a very limited number of models can actually be used to generate reliable predictions [3]. One of the reasons for this is the aforementioned lack of uniformly measured data for biodegradation of large numbers of compounds. Hence it is essential that close co-operation be established between experimentalists and modellers. What is needed is a multi-disciplinary approach to the problem.

2. Major Conclusions and Recommendations

On the basis of the general observations given above, the following set of conclusions and recommendations were derived:

- The degradability of a substance depends on its structure and physical form, the time that has been available for acclimation, and the environmental conditions. Importantly, these latter factors can be just as important as structure in determining the actual degradation of a chemical. Up until now, the emphasis in deriving predictive models for biodegradation has been on predicting the outcome of laboratory tests that simulate the environmental situation to a varying extent. It was therefore concluded that, given the limitations in the available data from laboratory and field biodegradation tests and limitations in our understanding of how to represent the many factors that effect biodegradation, there is a limit to what can realistically be done with this data to develop Quantitative Structure Biodegradation Relationships (QSBRs). However, if the biodegradation data are carefully chosen, it should be possible to develop good QSBRs for predicting if a chemical is completely biodegradable. The appropriate data set for developing this type of (Q)SBR will not just be pass/fail data for a given test but rather an evaluation of biodegradation data from different tests to determine if the substance is biodegradable. Another option is to use the data from pass/fail of a test to develop a (Q)SBR. However, it should be recognised that such a (Q)SBR will only predict the results of this test.
- QSBRs for predicting biodegradation rates, especially for a broad range of substances are going to be more difficult to develop for several reasons. First, the biodegradation rate is effected by a large number of confounding factors and methods do not exist to represent the effects of all these factors on the rate of biodegradation. However, as a better understanding of these confounding factors and the results of realistic test methods become available for more substances, then it will become possible to develop a QSBR to predict the biodegradation rate for a wide range of substances. Given the current limited possibilities of quantitatively predicting rates of biodegradation, it is recommended that one uses the terminology "SBR" instead of "QSBR".
- Testing of biodegradability is often performed in the laboratory. Often test conditions are prescribed that do not resemble the situation in the field. For instance; typical biodegradation testing is performed in simulation tests that use high concentrations of test substance and high concentrations of (adapted) inocculum. In the field however, much lower concentrations of both test substance and adapted microflora can be found. It was therefore recommended that biodegradation testing be performed in simulated laboratory settings that resemble as closely as possible to the actual field situation. Emphasis should be placed upon testing chemicals under the environmental conditions in which they are expected to reside (as a typical example: testing of highly hydrophobic chemicals in anaerobic environments, in contrast to hydrophilic non-volatile substances that are preferably tested in aerobic aqueous environments). This will not only facilitate extrapolation of laboratory test results to realistic conditions, but will also assist in the development of superior QSBRs capable to forecast what will happen in the 'real' environment.
- A major drawback in the development of Quantitative Structure Biodegradation Relationships (QSBRs) is the lack of uniformly measured data for biodegradation of large numbers of compounds (statistically sound models can never be better than the underlying data!). It was concluded that this is probably due to differences in objectives between scientists who actually perform biodegradation testing (major

aim: to collect a minimum set of data that will enable to judge whether a specific chemical is yes/no biodegradable) and scientists trying to develop QSBRs (major aim: collect as many data as possible for large numbers of chemicals). It is recommended that closer co-operation between the two groups of scientists needs to be established.

- In view of the consequences in the field, the major challenge when deriving QSBRs should be the identification of chemicals that are non-degradable under field conditions, instead of aiming at predicting degradability. It was concluded that due to insufficient knowledge of the actual degradation process, non-degradable compounds may often be classified as being degradable.

- In order to be able to derive reliable and broadly applicable QSBRs it is essential that knowledge of the biochemical transformation pathways that chemicals undergo in the field, is carefully taken into consideration. On the basis of knowledge of the biochemical pathways, rules are to be derived along which transformation takes place. Essential in this approach is the determination of the rate limiting step within specific pathways. At the moment this is possible only to a limited extent. It is therefore concluded that the prediction of biodegradation can be greatly improved by inclusion of the time scales during which the different processes take place. It was agreed that, due to natural selection of pathways for the metabolism of naturally occurring chemicals, there is not an infinite number of biochemical transformation pathways, as might be feared on forehand. From this it was recommended to further proceed towards mapping the most important catabolic pathways. Furthermore it was recommended that special attention needs to be paid to the possible formation of dead-end products (formation of persistent intermediates). Formation of intermediates that are far more toxic than the parent compounds cannot be excluded. It was therefore recommended that during biodegradability testing, the formation of intermediates and their toxicities be taken into consideration: QSBR models should be combined with toxicity data or predictions from QSAR models for toxicity. Thus, the toxicity of both the compound of interest and its metabolites could for instance be used as one of the descriptors in QSBR modelling.

- An approach that is often used in deriving QSBRs is related to the finding that to a certain extent a clear distinction can be made between structural fragments that enhance biodegradation (biophores, or molecular fragments that are recognised by microbes and subsequently transformed) and fragments that retard the process. This distinction was recognised before by experts in the field of biodegradation and was used to develop an expert system for predicting biodegradation. Computerised systems have now been developed that are based on structural fragments shown to retard or enhance biodegradation, yielding the probability that a given compound is degradable. By comparison with similar systems that have been developed for predicting toxicity, carcinogenicity and some physico-chemical properties, it was concluded that in principle there is no reason to treat biodegradation in a different manner.

- One of the conclusions with regard to model development was that at the moment the development of QSBRs is not hampered by a lack of well-defined descriptors and statistically based modelling techniques.

- It was concluded that knowledge on the biochemical transformation pathways can be achieved only by means of a multi-disciplinary approach. It was therefore recommended that closer co-operation be initiated between scientists in several distinct fields, like for instance biochemistry, quantum-chemistry, microbiology, analytical chemistry, statistics, etc.

- Finally it was recommended that the present question about environmental degradation "will a chemical biodegrade fast or slow?" should be formulated as "will a chemical biodegrade fast or slow under the following set of (field) conditions: (a) physical, (b) chemical, (c) biological, etc...?".

As a follow-up on the NATO Advances Research Workshop, further co-operation between the participants was initiated aimed at dealing with some of the major drawbacks that were identified. One of the major problems, which is the lack of uniformly measured biodegradation data, will be dealt with by means of exchange of databases available among the participants. The data will be processed according to the procedures presented during the workshop. Subsequently the results obtained (new QSBR-models) will be validated, exchanged and compared. This is expected to lead to additional recommendations on the use and application of QSBRs.

3. References

1. European Commission (1996) EU Technical Guidance Documents for risk assessment of new and existing substances, European Commission, Luxembourg.
2. Cowan, C.E., Federle, T.W., Larson, R.J., and T.C. Feijtel (1995) Impact of biodegradation test methods on the development and applicability of biodegradation QSARs, in W.J.G.M. Peijnenburg and W. Karcher (eds.), *Proceedings of the Workshop "Quantitative Structure Activity Relationships for Biodegradation"*, National Institute of Public Health and Environmental Protection (RIVM), Bilthoven, The Netherlands, report no. 719101021, pp. 104-115.
3. Langenberg, J.H., Peijnenburg, W.J.G.M., and Rorije, E. (1996) On the usefulness and reliability of existing QSBRs for risk assessment and priority setting, *SAR QSAR Environ. Res.* **5**, 1-16.

BIODEGRADABILITY OF XENOBIOTIC ORGANIC COMPOUNDS DEPENDS ON THEIR CHEMICAL STRUCTURE AND EFFICIENTLY CONTROLLED, AND PRODUCTIVE BIOCHEMICAL REACTION MECHANISMS

R.-M. WITTICH
Division of Microbiology, GBF - National Research Centre for Biotechnology, Mascheroder Weg 1, D-38124 Braunschweig, GERMANY

1. Introduction

Industrial chemicals like halogenated, sulphonated and nitrated aliphatics and aromatics, many of which represent xenobiotics, persist in the biosphere and need to be eliminated from the environment once they have entered as pesticides or other technical end products, industrial effluents, or unintentionally through accidents. Although one can find numerous articles in the literature on the biodegradation of such compounds, there is only little information on the biodegradability of most of the chemical structures synthesized up to now. The establishment of predictive models is an urgent need but this faces fundamental problems. The first will be the definition of biodegradability in terms of transformation and/or mineralization, and the biological processes involved, and the second is the goal one wishes to attain.

It has to be stated at the outset that, in principle, every chemical compound irrespective of its particular structure and dimension is open to attack, e.g. by fungal peroxidases. Such a highly unspecific oxidation by oxygen radicals, however, produces relatively small amounts (ca. 5%) of CO_2, in contrast to often more than 60% in case of specialized mineralizing bacteria, and mainly repolymerizes the oxidized organic matter and thus contributes to the formation of so-called bound residues. On the other hand, many chemical structures will be fortuitously transformed by fungi and bacteria able to degrade or mineralize compounds of similar structure, sometimes to dead-end products which are toxic and prevent further degradation.

It is essential to decide, what we really wish to predict. Is it the biodegradability by fungi or bacteria, or is it the degradation of compounds by bacteria thereby deriving energy, which would be the best case because of its self-maintaining property? The latter is obviously preferable but we have to accept that this process only works in a satisfactory, energy-generating way through a multi-dimensional regulation of degradative pathways enzyme activities and gene expression. This is not very well understood in many cases and will require much more fundamental research.

Furthermore, other physiological properties are important which determine the quality of biodegradabilty, like uptake systems, which are also regulated, and the concomitant production of biosurfactants, which facilitates degradation and enhances the

7

W. J. G. M. Peijnenburg and J. Damborský (eds.), Biodegradability Prediction, 7–16.
© 1996 *Kluwer Academic Publishers.*

bioavailability of potentially xenobiotic carbon and energy sources.

2. General Remarks

Irrespective of the class of chemical compounds or of a well-defined single compound we wish to consider for our biodegradation studies, or we are having in mind in terms of the prediction of their biodegradability, we have to make some more fundamental reflections with regard to a definition of biodegradation.

Biodegradation can take place somewhere in the natural or polluted environment, or in a biotechnological process, the latter being much more regulated and optimized for a final goal: a product which may be a single organic compound of higher value, or e.g. just CO_2, H_2O, and biomass from the complete aerobic mineralization of compounds. Biodegradation, however, demands also a physiological environment in which it can take place. Biodegradation of natural compounds in extreme environments also may occur, but reports on biodegradation of xenobiotics under such circumstances are negligible. Such conditions like high temperature ($>40°C$), extremely low or high ionic strength, very low or high pH - neutral between 6 and 8 is well tolerated, low oxygen pressure in aerobic processes, too low or high substrate concentrations, together with low bioavailability of substrates for the generation of energy and carbon for cell growth will prevent or slow down biodegradation and cause the accumulation of (xeno-biotic) compounds. The persistence and recalcitrance against microbial degradation of real xenobiotics, however, is caused predominantly by substitution of hydrogen atoms of natural structures by sulpho and nitro groups as well as halogens which require modified enzymes for their removal in order to make the carbon skeleton accessible for the microbes' central metabolism. This is of importance insofar as it determines the final biotechnological process which has to be applied or developed for its removal from the environment or from technical facilities. Prior to such application, however, the biological catalyst for the transformation of the target compound(s) has to be identified by selection from databases and other sources, or has to be isolated from the natural or technical environment. Probably, genetic engineering will provide hybrid organisms with new degradative capabilities.

2.1. GENERAL DEFINITIONS

First, however, we should define the term biodegradation, herewith disregarding that both bioabsorption and bioaccumulation are playing an important role in the depletion of compounds from natural environments and technical installations. Biodegradation of a compound in an aerobic environment e.g. may happen as a simple, one-step bio-transformation by a biocatalyst (an enzyme), rendering the parent structure a new one by removing or altering substituents, but without making relevant changes to the carbon backbone. This may happen by the action of an extracellular enzyme in an aqueous solution or humid solid (soil) system, or by the intracellular enzyme system of an organism. In the latter case the parent compound has to enter the cell through a complex cell wall and membrane system (several enzymes are localized between the cell wall and the membrane(s) or are found in the membrane(s)). This will be by passive diffusion of predominantly non-polar substances, including alcohols and phenolic compounds. More polar and, especially, dissociated compounds like aliphatic and aromatic acids need a transport system for uptake. These transport proteins are highly specific for single compounds and often exclude structural isomers or analogues. There are, however, indications that the same transport system is utilized for the uptake of e.g. aromatic carboxylic acids and the structurally analogous sulphonated aromatic compound. The presence or absence of transport systems also may have an important influence on the toxicity of the target compound onto the degrading organism because the specificity as well as the activity of the respective transport protein which, up to a certain extent, regulates the internal concentration. The latter, of cause, depends also of the compounds' reexcretion rate or its further fate inside the cell by subsequent catabolic enzyme reactions.

These then will make up a degradative sequence comprised of a set of more or less specific reactions, leading to a much more transformed (end) product of higher oxidized and/or reduced carbon content than that of the parent molecule. These products of enzymatic reactions, at the moment, may still be the result of fortuitous biotransformations, a more accidental event, as many enzymes exhibit a relatively broad substrate range. Such reaction sequence will stop when there is no enzyme present in the cell which is capable to perform any further transformation. The aforementioned reactions, on the other hand, will need a somehow generation of energy as the active cell needs for its self-maintenance. Consequently, there must another process take place to derive this energy from the metabolism of other compounds, thus driving the so-called co-metabolism of our target compound. Its partial degradation in several cases also may generate fragments of the former carbon skeleton which then are channeled directly into the cells' central metabolism for further energy formation; other fragments will be further transformed and probably excreted later or won't be trans-formed and accumulated and excreted on this stage. Consequently, such biological process is not complete in terms of the entire mineralization of the target compound we would like to take place because during complete aerobic mineralization ca. 70% of the initial carbon is transformed into CO_2 and ca. 30% into the growing biomass, thereby also releasing xenobiotic substituents such as nitro or sulpho groups, or halogen substituents as the respective anions together with protons. In order to achieve such a self-maintaining process which supplies necessary energy for the growth of the organisms the presence of a complete degradative pathway features a prerequisite.

3. Fungal Oxidations

3.1. OXIDATIONS BY PEROXIDASES

Many white-rot basidiomycetes, predominantly under N-limitation, produce extracellular lignin-type and manganese-dependent peroxidases as well as constitutively-induced laccases (also known as phenol oxidase). Whereas peroxidases require H_2O_2 for activity laccases utilize activated molecular oxygen. A wide range of ((halo)aromatc) compounds is reported to be degraded unspecificly by these oxidating and also ring cleaving enzymes under release of relatively small amounts of CO_2. A decrease of toxicity of target compounds comes along with concomitant dehalogenation and/or polymerisation; there is little information on the structures of formed polymers. Low-molecular weight monomers obtained from such oxidations are utilized for central metabolic and anabolic processes.

Starting from e.g. chlorophenols the biosynthesis of chlorinated dioxins has been reported. This reaction may be a detoxifying one only for fungi and procaryots. The class of haloperoxidases are utilized for regiospecific halogenations in organic syntheses; the respective fungal enzyme, however, is a P-450 oxidase.

3.2. OXIDATIONS BY P-450 SYSTEMS

Fungal and bacterial hemoprotein cytochrome P-450 enzymes in the presence of molecular oxygen oxidize numerous aliphatic and aromatic compounds, thereby exhibiting some kind of broad substrate specificity depending on the subclass of the enzyme. Although P-450 enzymes are capable of catalyzing reductions most of the P-450 oxidations proceed with the stoichiometrical characteristic of a monooxygenase reaction, requiring NAD(P)H. Subsequent hydroxylations of their substrate(s) have been reported.

3.3. PATHWAYS FOR COMPLETE DEGRADATIONS

Fungi possess, of course, also complete pathways for the mineralization of many natural low-molecular weight compounds, e.g. for degradation of anabolic intermediates such as aromatic amino acids. Dioxygenases as well as hydrolases, lyases, and others have been described to make up a productive sequence of biocatalysts. By evolutionary processes capabilities for the mineralization of several xenobiotics were established, too. Although the regulations of these pathways at the moment are not as well under-stood as those of bacteria, they also seem well regulated by inducers on the gene level. The end products of these pathways allow further growth of the organisms and supply energy and carbon for the biological formation and operation of the biocatalysts, including the above-mentioned peroxidases and P-450 enzyme systems.

4. Degradation by Bacteria

Under this heading, anoxic/anaerobic biocatalytic processes won't be regarded in order to keep the present discussion at a minimum and the focus will be only on several examples of the aerobic degradation of several selected (halo)aromatics.

Although above P-450 oxidations of xenobiotics are also found in bacteria, for this group of organisms numerous pathways for the mineralization of natural as well as xenobiotic compounds have been reported. These pathways are comprised of sets of individual enzymes which catalyze different subsequent reactions, channeling the target compound into the central metabolism (Krebs cycle) for energy generation and supply of carbon for anabolic processes. The genes coding for these catabolic enzyme systems in most cases are found clustered on the bacterial chromosome or on a catabolic plasmid. Expression of the cluster of the degradative sequence in many cases is triggered by the substrate of the first enzyme, e.g. the target compound, or by an intermediate of the pathway, which may differ from organism to organism and sometimes depends on the bacterial species for the induction of, in principle, identical pathways. Full expression then results in a ca. 20-fold to about 500-fold increase in activity of the degrading enzyme system. An important prerequisit for the function of gene expression is the necessity of correct configuration of the inducer which has to fit properly for operation. If this is not the case, gene expression occurs on a reduced level or will be completely prevented. Since parallel pathways are found in a single microbe, e.g. for catechol, hydroxyhydroquinone, and for gentisate which are central intermediates in the degradation of aromatic compounds, as well as isoenzymes for the degradation of their halogenated derivatives, misrouting often is a problem which results in the accumulation of dead-end products that are detrimental to the organism and end up in the decease of the microbe caused by intracellular poisoning of the biocatalysts or of other functional units of the cell. The same may be true in the case of the unproductive catabolism of a structural analogue which is turned over with reduced activity to a step on which an intermediate accumulates intra- and later extracellularly. A similar problem will come up in case of an incomplete pathway.

On the other hand, a structural analogue can function as an inducer and finally lead to the fortuitous induction of a complete but futile set of enzymes.

4.1. EXAMPLES OF PATHWAYS FOR DEGRADATION OF HALOAROMATICS

In Figure 1 a couple of haloaromatic compounds are shown. Most of the nonhalo-genated parent structures are mineralized through one or several highly specific enzymatic reactions to, and then via catechol by so-called type I enzymes. The corresponding haloaromatics are, if at all, mineralized by type II isoenzymes which predominantly depends on the particular substitution pattern of the single isomers of the target compounds. The initial attack takes place by the action of a dioxygenase of high regioselectivity, but often of relaxed substrate specificity. Although a (halo)benzene dioxygenase does not attack neither (halo)phenols nor halogenated biphenyls, dibenzodioxins, and dibenzofurans, the (chloro)biphenyl 2,3-dioxygenase is capable to attack the two latter compounds at lateral positions, forming *cis*-dihydrodiols as in case of biphenyl. Unfortunately, these lateral dihydrodiols, although rearomatizable to diols by a dehydrogenase, resist further breakdown because in the case of productive

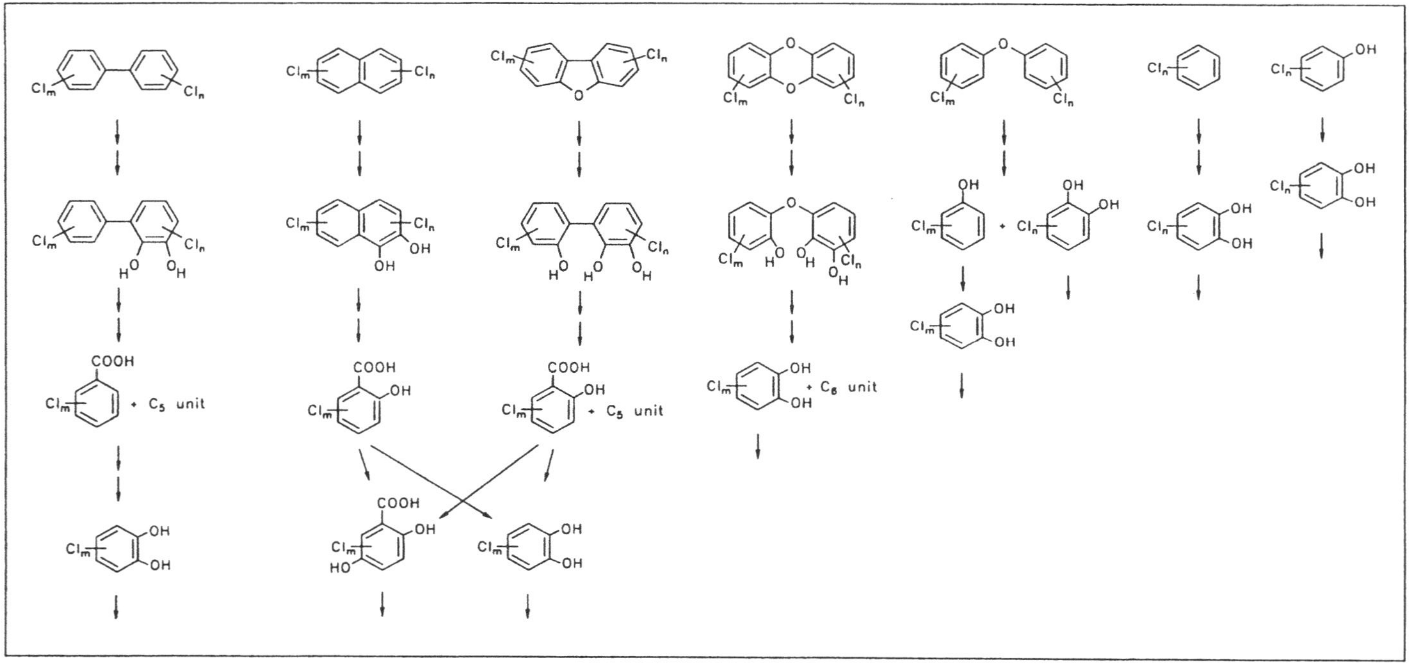

12

Figure 1. Converging pathways for aromatic compounds and their chlorinated (halogenated) derivatives

degradation of dibenzodioxins, dibenzofurans, and also of diphenyl ethers another (new) type of dioxygenase is necessary which attacks at the ether bridge. This dioxin dioxygenase furnishes instable hemiacetals which spontaneously decay to tri-hydroxylated aromatics or, in case of the diphenyl ether, directly into phenol and catechol. The initial attack of the dioxin dioxygenase onto the 3 different mono-halogenated diaryl ethers leads to a couple of halogenated intermediates which are channeled into divergent and unproductive co-oxidizing sequences because no operating pathways exist for so many different halogenated intermediates as shown in Figure 2, there is only one for non-halogenated dibenzodioxin and its metabolites. Furthermore, none of all of the examined diaryl ether co-oxidizing microbes possesses the necessary Type II enzymes for further breakdown of haloaromatics.

The attack of (halo)benzene dioxygenase onto halogenated benzenes always generates a single intermediate. Furthermore, the halobenzene dioxygenase is capable to abstract dioxygenolytically a first chlorine in case of 1,2,4,5-tetrachlorobenzene (figure 3), but not in case of other isomers; several of them cannot be attacked at all as is demonstrated in Figure 4.

The non-accessability of these 3 (4) congeners, all of them exhibiting the 1,3,5-substitution pattern, by (halo)benzene dioxygenase defies calculations of electron densities and other parameters, and also has to take into account the sterical hindrance by bulky substituents; QSAR/QSBR examinations would be very useful in these cases. Modelling would be also of interest in terms of the prediction of biodegradability of all of the chloro(halo)benzenes shown in Figure 5.

This is of particular interest because there is only little information on the degradation of 1,2,3-trichloro- and 1,2,3,4-tetrachlorobenzene in the literature. All chlorobenzene congeners can be reductively dehalogenated under anaerobic/ methanogenic conditions, functioning as an electron acceptor for the respective microflora of

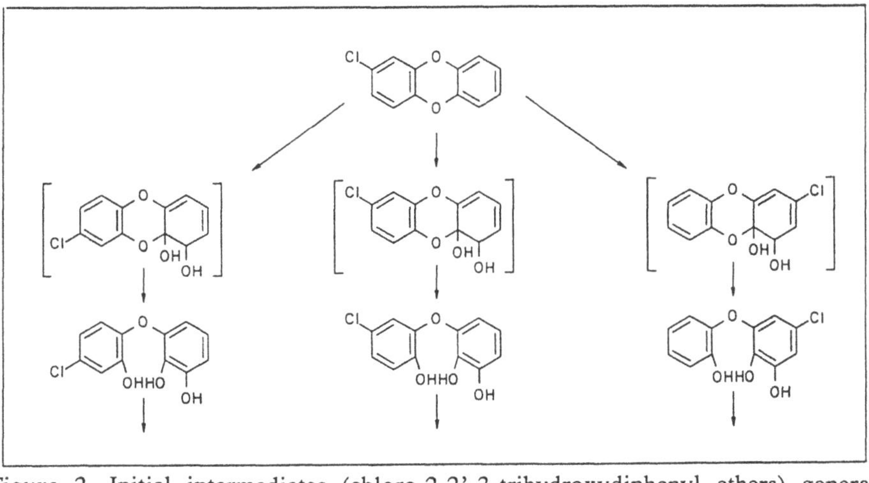

Figure 2. Initial intermediates (chloro-2,2',3-trihydroxydiphenyl ethers) generated through dibenzodioxin dioxygenase attack onto 2-chlorodibenzo-p-dioxin.

14

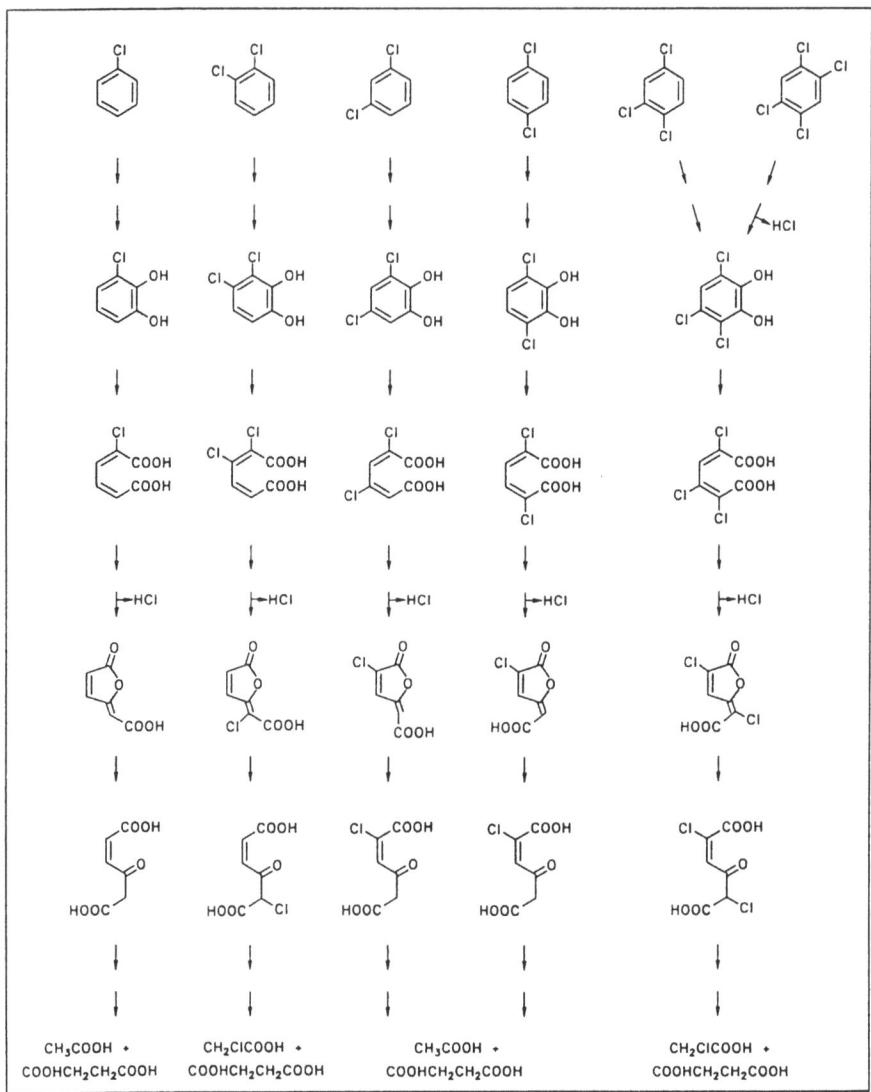

Figure 3. Pathways for the aerobic mineralization of chlorobenzenes by *Burkholderia* sp. strains PS12 and PS14.

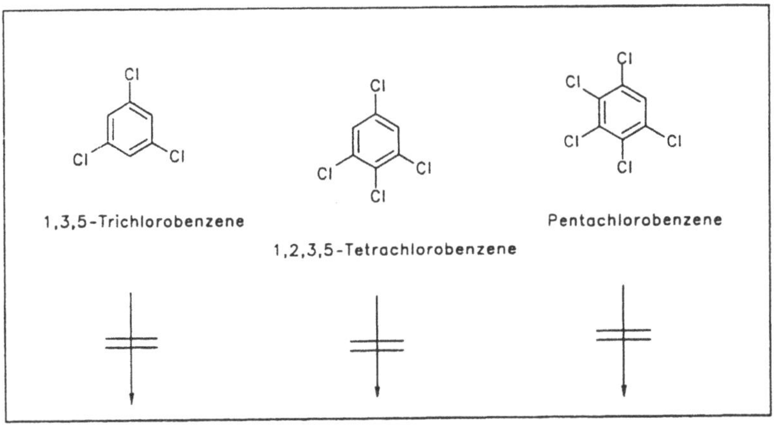

Figure 4. Chlorobenzenes resisting aerobic bacterial attack. Hexachlorobenzene is not shown here because it offers no free (non-halogenated) site for a dioxygenase reaction.

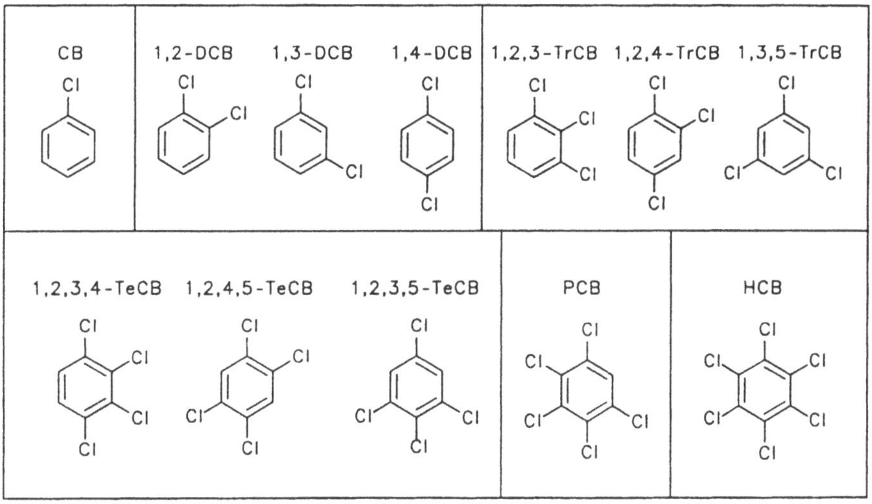

Figure 5. Substitution pattern of chlorobenzenes.

such environments. These two isomers and especially the latter are of important environmental and technical concern. In aerobic environments, however, the mineralization of only those chloro-benzenes shown in Figure 3 by pure bacterial cultures has been reported, and no single organism or mixed culture has been found which is capable to degrade the 1,2,3-trichlorobenzene. The first reason for this observation may be the fact, that 3,4,5-trichlorocatechol will be the central intermediate. This is extremely critical for degradation because chlorocatechols of the 4,5-dichloro substitution pattern are known as strong inhibitors of the type II chlorocatechol 1,2-dioxygenase and this should also be true for the tetrachlorocatechol in 1,2,3,4-tetrachlorobenzene degradation. The next critical step for the 4,5-dihalogenated catechols is found on the breakdown level of the maleyl acetates. The respective catalyst, chloromaleyl acetate reductase, is able to eliminate the chlorine from position 2, but not from position 3, adjacent to the keto group. Consequently, above 1,2,3-trichlorobenzene cannot be mineralized aerobically, although it can be initially attacked by chlorobenzene dioxygenase, a fact what might be confirmed by QSAR examinations.

Astonishingly, we recently succeeded in the isolation of a *Pseudomonas chlororaphis* strain capable of growth at the expense of 1,2,3,4-tetrachlorobenzene, but not of 1,2,3-trichlorobenzene. Probably we have to elucidate new mechanisms for the elimination of chlorine in the course of our future investigations.

5. Conclusions

The predictability of the biodegradability of a defined organic compound mainly depends on the biological catalytic system: in principle, peroxidase-activated oxygen attacks nearly all kinds of organics in contrast to P-450 enzymes which bind and convert their substrates much more specificly, making QSAR/QSBR calculations quite useful.

High enzyme specificity is also found for most of the degradative enzymes of many productive bacterial pathways for the complete mineralization of target compounds.

Most likely, the application of QSAR/QSBR onto complete pathways as a whole in terms of a black box system doesn't make much sense and the success of an improved modelling cannot be foreseen at this moment. QSAR application onto single reactions of a pathway, however, might be useful if one has in mind to use sets of single modules for the prediction. Similar modules are used in genetic pathway engineering on a molecular' level in the laboratory. However, nature also utilizes these instruments: confronting microbes with new compounds (real xenobiotics) in many cases will lead to the productive rearrangement of genetic information through spontaneous mutations and the uptake and integration of suitable genetic elements which then have to be put under the control of a regulating system. All these events in a whole are contributing to an evolution process developing itself along a time axis.

BIODEGRADABILITY TESTING OF XENOBIOTICS

P. PITTER, V. SÝKORA
Department of Water Technology and Environmental Engineering,
Prague Institute of Chemical Technology,
166 28 Prague, CZECH REPUBLIC

1. Abstract

An overview is given of the main aspects related to the biodegradability of chemicals. As such, two approaches to the evaluation of biodegradability are distinguished: a microbial approach (usually based on experiments with pure bacterial cultures at optimum conditions)and an environmental approach (based on experiments with mixed cultures grown under conditions approaching field conditions). Several types of biodegradability tests are reviewed, and testing strategies are discussed.

As a basis for developing Quantitative Structure Activity Relationships for biodegradation, detailed results of biodegradation testing are reported for several classes of chemicals.

2. Introduction - General Aspects

Degradation can in principle be complete or partial. **Complete degradation** denotes the transformation of the organic compound to carbon dioxide and water, with a concomitant formation of fresh microbial biomass. **Partial degradation** involves either a partial decomposition of the molecule (e.g. degradation of an aliphatic chain up to a quarternary carbon atom) or the formation of new, more complex compounds which are biologically stable (e.g. formation of polyphenols from simple phenols). Partial degradation is often associated with **primary degradation**. This term derives from the study of surfactants; in this field, primary degradation denotes a minimum transformation of a molecule which leads to the disappearance of some of its characteristic properties. For instance, a substantial shortening of the hydrophobic aliphatic chain in surfactants leads to a loss of surface activity and thereby to a loss of the foaming ability. The term primary degradation was coined in a period when detergent laws were being set up, and has been losing its dominance with the advent of laws on chemicals. Partial degradation cannot be considered environmentally sufficient since the ensuing stable metabolites may be accumulated in the environment [1,2,3].

Biodegradability can be assessed in terms of its purpose and in terms of its conditions. Classification of chemicals into biologically degradable and nondegradable compounds is both relative and inaccurate. One should in each case specify the conditions under which no biological degradation was (or was not) proved. Seemingly, biologically nondegradable compounds are usually in fact merely slowly degradable, or

17

W. J. G. M. Peijnenburg and J. Damborský (eds.), Biodegradability Prediction, 17–26.

their degradability can be enhanced by selecting suitable microorganisms. Basically, there are no substances which would be entirely biologically nondegradable. The degradation may be so slow that, in terms of self-purification in streams, biological treatment of waste waters and the behaviour of chemicals in dumps, is virtually negligible (e.g. humic substances, lignin sulfonates, synthetic polymers).

There are two approaches to the evaluation of biodegradability: microbiological and environmental. The **microbiological approach** is usually based on experiments with pure bacterial cultures for which optimum conditions are sought and whose metabolism is studied. These conditions, however, differ from those prevailing in natural environment and the ensuing data therefore cannot be used directly for predicting the actual behaviour of chemicals in the environment. The degradability data resulting from this approach are usually overestimated. However, the results of these experiments have their practical importance for biological removal (bioremediation) of accidental pollution of ground waters or soil, which is usually performed with pure bacterial cultures. The **environmental approach** focuses on the biodegradability of chemicals in water and soil which are recipients of liquid and solid wastes. The environmental evaluation is based on experiments with mixed microbial cultures collected in the particular recipient location or in a waste water treatment plant, or with mixed cultures grown in the laboratory under conditions simulating biological remediation or self-purification in surface waters. The results of this assessment are therefore much nearer to actual field conditions. For this reason, most standardized methods for biodegradability determination are based on the environmental approach [3,4].

3. General Aspects of Biodegradability Testing

One of the decisive aspects in the classification of chemical is whether or not they are biodegradable. This classification is important both environmentally and for legislature since it aids in distinguishing chemicals which pose no environmental problems to the environment from those which merit special attention and whose production and/or application in various products could be curtailed or prohibited.

A group of chemicals which poses essentially no problems are so-called **readily biodegradable substances**. These should meet the following conditions:
(I) they can serve as the only carbon and energy source for micoorganisms;
(ii) they can be completely mineralized to carbon dioxide and water;
(iii) they do not require any complex adaptation and selection of microbial degraders;
(iv) the rate of their degradation is comparable with the rate of degradation of organic substances present in sewage waste waters;
(v) the degradation has to proceed rapidly and to an extent that eliminates the appearance of undesirable environmental effects.

Tests for ready biodegradability simulate natural conditions with a relatively low concentration of both the tested substance and the microorganisms, and they involve only partially the effect of adaptation. When the test has demonstrated a ready biodegradability, it can be safely assumed that the compound will be readily degraded also under natural conditions. Further testing is then unnecessary. On the other hand, negative results of tests for ready biodegradability do not necessarily imply that the compound will be biologically resistant under various environmental conditions.

The classification of biodegradability therefore includes a category of potentially biodegradable compounds. **Tests for potential degradability** aim at determining whether degradation is in principle possible. Tests for potential biodegradability simulate to a certain extent the conditions in biological wastewater treatment plants and natural conditions, and hold for the situation when the particular environment already possesses prerequisites for development of microbial degraders of the given substance.

A **Testing strategy** requires a certain hierarchy of procedures. The basic question is whether or not the substance is biodegradable. An assay for ready biodegradability is carried out by a number of tests. It should be noted that a positive or negative result of a single test for ready biodegradability is not sufficient for an overall classification of the substance. The tests usually involve a combination of BOD assay or carbon dioxide production with determination of DOC loss. Unless ready biodegradability was proved by at least these two tests, the next step involves tests for potential degradability. If the results of the tests are negative, there follow simulation tests in which wastewater treatment technology is simulated under laboratory conditions, most often with activated sludge. When even these trials are insufficient to reach the desirable degradability limit, the substance can be denoted as **poorly biodegradable (resistant)**. The definition of these conditions is very important. Degradation of substances classified as poorly biodegradable in the environment has been shown to depend largely on microorganisms with long generation times [5,6]. The higher per cent biodegradation found in experiments with biological columns and biofilm reactors is largely due to the fact that the biofilms are prone to become enriched with longer-generation-time microorganisms. Irrespective of this fact, however, it is clear that substances that require selection of microorganisms or an application of less common technological processes for wastewater treatment cannot be considered readily biodegradable.

These variations of **simulation tests** represent basically the termination of the environmental approach to biodegradability testing. If necessary, the microbiological approach follows which provides information about the microbial species participating in biodegradation and involves isolation of pure cultures and examination of their metabolism. This, however, represents an extensive research work which requires a highly individual attitude even though it has certain general features. Let us stress again that this microbiological approach, though important for revealing the causes of the apparent resistance, does not permit a prediction of the behaviour of the chemicals in the environment except in cases of biological remediation of ground waters or minerals.

Relatively little attention has so far been paid to **biodegradability under anaerobic conditions**. Most organic compounds are better transfornmed under aerobic conditions but this does not hold for all chemicals. For instance, polyhalogenated compounds including polychlorinated biphenyls (PCB) are better dehalogenated under anaerobic conditions [7]. Assessment of anaerobic biodegradability has a special importance since the remediation of wastewater treatment sludges is usually performed under anaerobic conditions and some industrial wastewaters are also successfully treated by anaerobic processes. Under natural conditions, anaerobic degradation may occur in river sediments and in soil layers, and in deeper horizons of ground waters. Anaerobic processes have their place in testing strategy, especially in the case of chemicals resistant under aerobic conditions.

Effort has been made to treat the experimental biodegradability data by **correlation analysis.** Such correlations enable to predict biodegradation rate values or to estimate readily or non-readily biodegradation of experimentally non-tested organic compounds. Understanding QSAR between chemical structure and environmental stability provides insight into breakdown mechanisms and pathways in the environment. Such understanding allows organic chemists to develop materials having optimal biotransformation potential. It appears that the pursuit of this approach might be very fruitful [3,8,9].

For this purposes it is necessary to elect a series of organic compounds with different type and position of substituents, and then translate the molecular structure into different descriptors. On the other hand it is necessary to obtain experimental data about rate and degree of degradation. For this purpose different possibilities can be taken into account (e.g. first or second order rate constants, constants from the Michaelis equation, specific rate of biodegradation expressed as the amount of the substrate removed per gram of the initial amount of the microbial culture per hour, time for reaching the same percentage of biodegradability of standard compounds, biochemical oxygen demand (BOD) as the percent of theoretical oxygen demand, degradation half-live). From the **environmental point of view,** there are several methods (which use mixed bacterial cultures) available to obtain above-mentioned parameters [4]. Each of the biodegradability tests has limitations that effect the data available for QSAR approach. Screening biodegradability data must be very carefully evaluated. The problem of acclimated or unacclimated microorganisms or application of pure microbial cultures (**microbiological point of view**) must also be taken into consideration. Degradation rates can be safely compared only when all experimental conditions, in particular inoculum history, are the same. Kinetic tests, which are common in water treatment technology, make it possible to obtain reasonable kinetic parameters which are relatively relevant for the environmental conditions under study [10].

The OECD methods for determination of biodegradability of organic substances allow a wide range of usage of microbial inoculum concerning both its origin and quantity. Inoculation is acceptable by surface water, by treatment plant effluent and by activated sludge. The amount of inoculum varies from millilitres to dozens millilitres per liter of medium, so that the quantity of microorganisms in the inoculum can differ in order of magnitude. This possibility of a wide usage of inoculum origin and quantity can have negative consequences because it influences not only the overall course but also the degree of biological degradation. In these cases the comparison of inter-laboratory test results is troublesome. One laboratory can classify the substance tested as readily biodegradable but an other laboratory may classify the same chemical as being not readily biodegradable because of little inoculum usage with a low quantity of microorganisms. Data achieved in this way are hard to correlate because the real degradability of organic substances could be underestimated.

4. Some Results

Figures 1 and 2 summarize some biodegradation tests results (closed bottle test according to OECD 301D). The tested substances are sodium benzoate (recommended as a reference substance) and linear alkylbenzenesulfonate sodium salt (Marlon A390). Two surface water sources (SW1, SW2) and three biological sewage water treatment

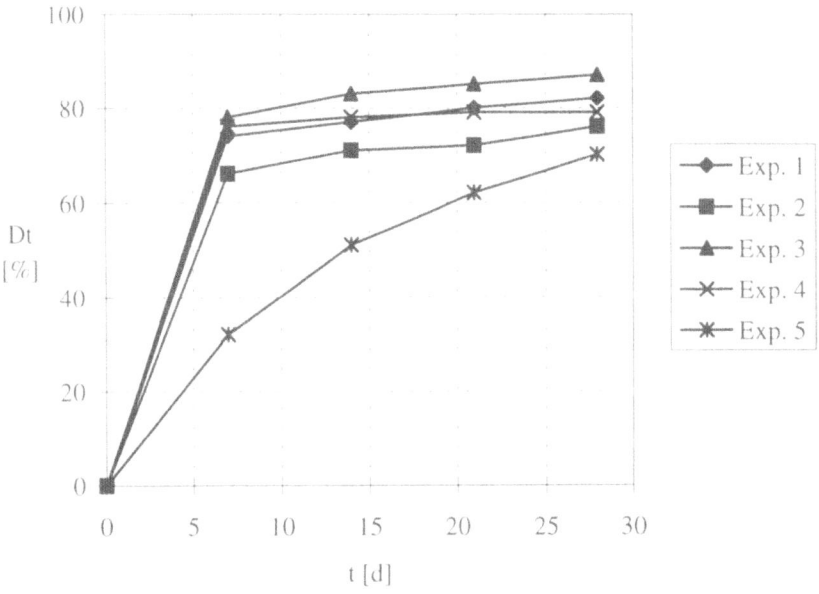

Figure 1. The biological degradation of sodium benzoate.

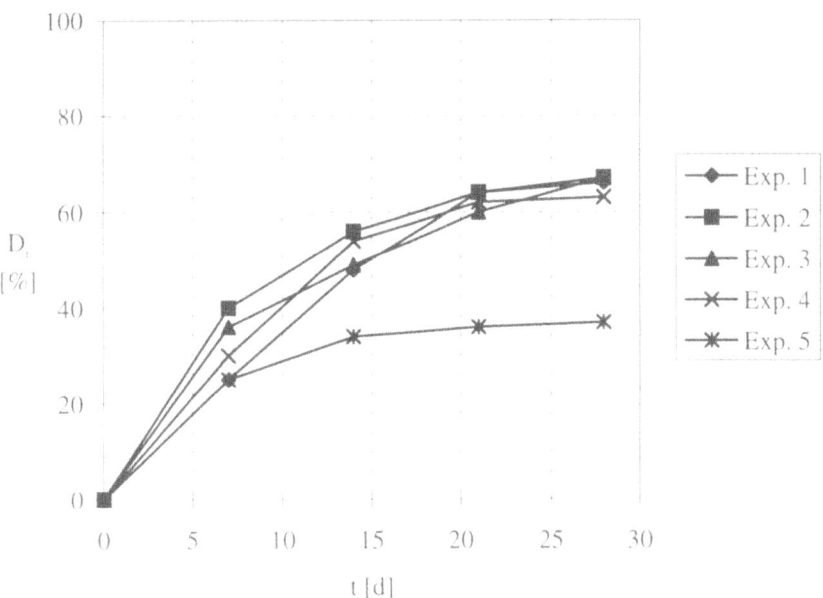

Figure 2. The biological degradation of linear alkylbenzenesulfonate sodium salt.

plant effluent sources (EF1, 2, 3) were used for inoculation. SW1 was an average river water while SW2 source was a surface water with low microbial population.

Experiment	1	2	3	4	5
Inoculum source	SW	EF	EF	EF	SW
Quantity [ml/l]	15	4	34	2	20

The course of the biochemical oxygen demand in the Figures 1 and 2 is expressed in TOD percentage (theoretical oxygen demand) according to OECD 301D. The substance is considered to be readily biodegradable if the BOD value after 28 days (D_{28}) is at least 60 % of TOD. This value was reached in all cases of sodium benzoate although degradation in experiment 5 was the slowest.

The results with SW2 inoculum, however, were entirely different from the others in experiments with linear alkylbenzenesulfonate degradation. Even after an incubation period of 28 days the value D_{28} did not exceed 40 % TOD and the limit value was practically reached. This means that only the laboratory using a low quantity of inoculum would classify the substance as not readily biodegradable, which does not agree to common known facts corresponding to the reality. So the results of improper inoculum type tests could negatively influence interpretation of QSAR relations and other correlations.

The appropriate quantity of inoculum could be defined either by the number of microorganism cells in unit volume or by endogenous respiration. The endogenous respiration of the inoculum is limited on the one hand by the value 1.5 mg/l in a 28 days experiment (not to exceed the respiration responsible for degradation of tested substance).

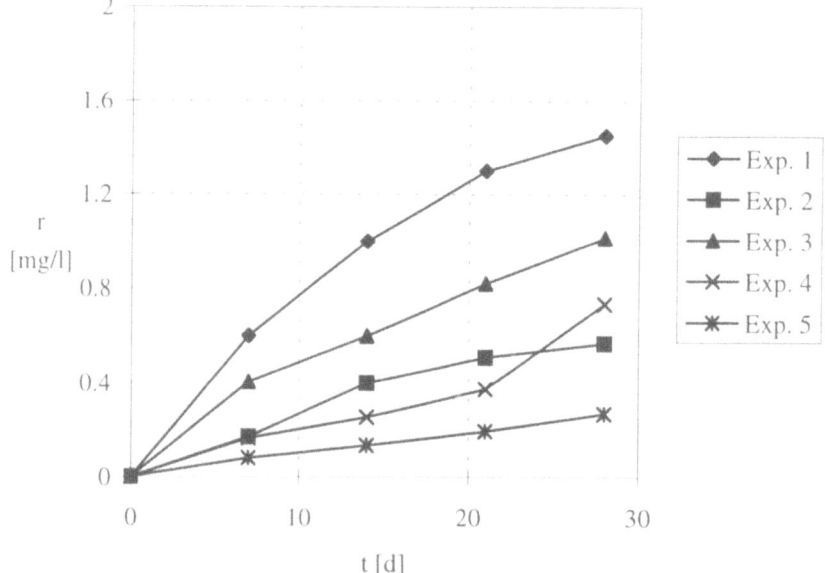

Figure 3. The endogenous respiration of the blank.

On the other hand, the inoculum must have a certain minimal endogenous respiration proportional to the number of living microorganism which is vital for biological degradation. The courses of endogenous respiration for five blank experiments are given in Figure 3. A blank experiment with the SW2 inoculum (exp. 5) had a very low endogenous respiration what was consequently shown on degradation of linear alkylbenzenesulfonate sodium salt (Marlon A390). It is evident that the endogenous respiration of the inoculum should not fall under 0.5 mg/l in a 28 days experiment.

TABLE 1 includes the maximum specific removal rates obtained by the above-mentioned method [10] of different aliphatic and aromatic organic compounds that are mutually comparable since they were obtained by means of an identical method using an adapted mixed culture (activated sludge). This microbial culture was cultivated semicontinuously on glucose and peptone and subsequently on the compound under study at a mean biomass retention time (MBRT) of approximately 5 days. This MBRT corresponds to biological treatment in the classical activated sludge wastewater treatment process. During the biodegradation experiment the substance tested was the sole source of organic carbon for the microbes of the culture. The initial concentration of the compounds tested corresponded to the oxygen equivalent (TOD - theoretical oxygen demand or COD - chemical oxygen demand) of 200 mg/l. The maximum specific removal rates are expressed as mg TOD(COD) of the compound removed per gram of the initial solids of the microbial culture per hour (q_{TOD}) and as μmol of the compound removed per gram of the initial solids of the microbial culture per hour ($q_{μmol}$) These values were already used for correlation between the structure of aromatic compounds and the rate of their biological degradation [11,12].

TABLE 1. Maximum specific rates q_{TOD} and $q_{μmol}$ of organic compounds removal (M - molecular weight, TOD - theoretical oxygen demand, COD - chemical oxygen demand)

Compound	M	TOD	$q_{TOD(COD)}$	$q_{μmol}$
	g/mol	g/g	mg/(g.h)	mmol/(g.h)
Aliphatic and cycloaliphatic compounds				
Acetic acid	60.05	1.07	50	778.2
1,4-Butanediol	90.12	1.95	40	227.6
1-Butanol	74.12	2.59	84	437.6
2-Butanol	74.12	2.59	55	286.5
ε-Caprolactame	113.16	2.12	63	262.6
α-Alanine	89.09	1.08	90	935.4
Allylamine	57.10	2.24	30	234.6
Chloroacetic acid	94.50	0.51	38	788.5
1-Cloropropionic acid	108.52	0.88	24	251.3
2-Cloropropionic acid	108.52	0.88	23	240.8
1,2-Cyclohexanediol	116.16	2.42	66	234.8
Cyclohexanol	100.16	2.72	28	102.8

Cyclohexanone	98.15	2.61	30	117.1
Cyclohexylamine	99.18	2.75	39	143.0
Cyclopentanol	86.13	2.60	55	245.6
Cyclopentanone	84.12	2.48	57	273.2
Diethyleneglycol	106.12	1.51	14	87.4
Dichloroacetic acid	128.94	0.25	18	558.4
1,1-Dichloropropionic acid	142.92	0.56	14	174.9
2,2-Dichloropropionic acid	142.92	0.56	14	174.9
Diethanolamine	105.14	1.52	11	68.8
Ethylene glycol	62.07	1.29	42	524.5
2-Ethoxyethanol	90.12	1.96	19	107.6
Ethanolamine	61.08	1.31	26	324.9
Ethylenediamine	60.01	1.33	22	275.6
Glucose	180.16	1.07	180	933.7
Glycerol	92.09	1.22	85	756.6
Lactic acid	90.08	1.07	73	757.4
4-Methylcyclohexanol	114.19	2.81	40	124.7
4-Methylcyclohexanone	112.17	2.71	63	207.2
2-Methyl-2-propanol	74.12	2.59	30	156.3
Nitrilotriacetic acid	191.14	0.75	8	55.8
Propionic acid	75.07	1.51	44	388.2
1-Propanol	60.01	2.40	71	493.0
2-Propanol	60.01	2.40	52	361.1
Tetrahydrofurfurylalcohol	102.13	2.04	40	192.0
Triethyleneglycol	150.17	1.60	28	116.5
Triethanolamine	149.19	1.61	18	74.9

Monosubstituted Benzene Derivatives				
Acetanilid	135.17	2.13	15	52.1
4-Aminoacetanilide	150.18	1.81	11	40.5
Aniline	93.13	2.41	19	84.7
Benzenesulfonic acid	158.18	1.57	11	44.3
Benzoic acid	122.12	1.97	86	357.5
Nitrobenzene	123.11	2.08	14	54.7
Phenol	94.11	2.38	80	357.2

Disubstituted Benzene Derivatives (ortho)				
2-Aminobenzoic acid	137.14	1.63	27	120.8
2-Aminophenol	109.13	1.91	21	100.7
2-Chloroaniline	127.57	1.63	11	52.9
2-Chlorophenol	128.56	1.62	14	67.2
o-Cresol	108.14	2.52	54	198.2
2-Nitrobenzaldehyde	151.12	1.70	14	54.7
2-Nitrobenzoic acid	167.12	1.44	20	83.3
2-Nitrophenol	139.11	1.50	14	67.1

2-Nitrotoluene	137.14	2.10	33	114.6
Phthalic acid	166.13	1.44	78	326.0
Pyrocatechol	110.11	1.89	56	269.1
Salicylic acid	138.12	1.62	95	424.6
o-Toluidine	107.16	2.54	15	55.1

Disubstituted Benzene Derivatives (meta)				
3-Aminobenzoic acid	137.14	2.63	7	19.4
3-Aminophenol	109.13	1.91	11	52.8
1,3-Benzenedisulfonic acid	238.23	1.01	4	16.6
3-Chloroaniline	127.57	1.63	6	28.9
3-Chlorophenol	128.56	1.62	32	153.6
m-Cresol	108.14	2.52	55	201.8
Isophthalic acid	166.13	1.45	76	315.5
3-Nitrobenzaldehyde	151.12	1.70	10	39.0
3-Nitrobenzoic acid	167.12	1.44	7	29.1
3-Nitrophenol	139.11	1.50	18	86.3
3-Nitrotoluene	137.14	2.10	21	72.9
Resorcinol	110.11	1.89	58	278.7
m-Toluidine	107.16	2.54	30	110.2

Disubstituted Benzene Derivatives (para)				
4-Aminobenzoic acid	137.14	2.63	13	36.0
4-Aminophenol	109.13	1.91	17	81.6
4-Chloroaniline	127.57	1.63	6	28.9
4-Chlorophenol	128.56	1.62	40	192.1
p-Cresol	108.14	2.52	55	201.8
Hydroquinone	110.11	1.89	54	259.5
4-Hydroxybenzoic acid	138.17	1.62	100	446.8
4-Nitrobenzaldehyde	151.12	1.70	14	54.7
4-Nitrobenzoic acid	167.12	1.44	20	83.3
4-Nitrophenol	139.11	1.50	16	76.7
4-Nitrotoluene	137.14	2.10	33	114.6
Sulfanilic acid	173.19	1.29	4	17.9
4-Toluenesulfonic acid	172.20	1.67	9	31.3
o-Toluidine	107.60	2.54	20	73.2

Trisubstituted Benzene Derivatives				
2-Chloro-4-nitrophenol	173.56	1.25	5	23.1
2,4-Diaminophenol	124.14	1.55	12	62.4
2,4-Dichlorophenol	163.00	1.18	11	57.2
2,5-Dihydroxybenzoic acid	154.12	1.35	80	384.5
2,3-Dimethylaniline	121.18	2.64	13	40.6
2,5-Dimethylaniline	121.18	2.64	4	12.5
3,4-Dimethylaniline	121.18	2.64	3	9.4

2,3-Dimethylphenol	122.17	2.62	35	109.3
2,4-Dimethylphenol	122.17	2.62	28	87.5
2,5-Dimethylphenol	122.17	2.62	11	34.4
2,6-Dimethylphenol	122.17	2.62	9	28.1
3,4-Dimethylphenol	122.17	2.62	14	43.7
3,5-Dimethylphenol	122.17	2.62	11	34.4
2,4-Dinitrophenol	184.11	1.22	6	26.8
Phloroglucinol	126.11	1.52	22	114.8
5-Sulfosalicylic acid	218.22	1.03	11	49.1

5. References

1. Painter, H.A., and King, E.F (1985) Biodegradation of Water-Soluble Compounds, *The Handbook of Environmental Chemistry,* Vol 2/Part C, Springer Verlag, Berlin, 87-120.

2. DeHenau, H. (1993) Biodegradation, *Handbook of Ecotoxicology,* Vol 1, Blackwell Scientific Publications, Oxford, 355-377.

3. Pitter, P., and Chudoba, J. (1990) *Biodegradability of Organic Substances in the Aquatic Environment,* CRC Press, Boca Raton, Florida.

4. OECD (1992) *Guidelines for Testing of Chemicals,* 301 A-F, 302 A-C, Paris.

5. Čech, J.S., and Chudoba, J. (1988) *Acta hydrochimica et hydrobiologica* **16**, 313-323.

6. Kobayashi, H., and Rittmann, B.E. (1982) *Environmental Science and Technology* **16**, 170A-183A.

7. Vogel, T.M., Criddle, C.S., and McCarty, P.L. (1987) *Environmental Science and Technology* **21**, 722-732.

8. Niemi, G.J., Veith, G.D., Regal, R.R., and Vaishnav, D.D. (1987) *Environmental Toxicology and Chemistry* **6**, 515-527.

9. Boethling, R.S., Howard, P.H., Meylan, W., Stiteler, W., Beauman, J., and Tirado, N. (1994) *Environmental Science and Technology* **28,** 459-465.

10. Pitter, P., and Sýkora, V. (1996) Biological Degradability Testing, *Environmental Xenobiotics* (M. Richardson, Ed.), Taylor & Francis, London (in press).

11. Pitter, P. (1984) Correlation between the structure of aromatic compounds and the rate of their biological degradation, *Collection Czechoslovak Chem. Comm.:n.* **49**, 2891-2896.

12. Pitter, P. (1985) Correlation of microbial degradation rates with the chemical structure, *Acta hydrochim. hydrobiol.* **13**, 453-460.

The META-CASETOX System

for the Prediction of the Toxic Hazard of Chemicals Deposited in the Environment.

G. Klopman
Chemistry Department
Case Western Reserve University, Cleveland, OH 44122 USA
and
MULTICASE Inc., Cleveland OH 44122 USA

Abstract

The operation and purpose of the META-CASETOX suite of computer programs is presented. CASETOX evaluates the toxic potential of organic molecules while META, outfitted with a biodegradation module evaluates the nature of the biodegradation products of organic molecules subjected to aerobic biodegradation. Together, they offer the ability to assess the toxic hazard posed by the disposal of organic molecules in the environment.

1.   Introduction

The widespread use of man-made chemicals by society has lead to many chemicals being released to the environment. The potential health and economic problems which can result from a chemical being deposited in the environment will depend on the persistence and toxicity of the parent compound and the metabolites it can be transformed into.   Since biodegradation of chemicals by microorganisms is such an important breakdown mechanism that can lead to detoxification and reduction of chemical accumulation in the environment, efforts are ongoing to uncover the relation between the structure of chemicals and their ability to biodegrade.   To determine the environmental fate of any chemical, the metabolic products of the parent compound must be examined so that the extent of degradation and the toxicity of the products can be assessed.

The risks associated with a chemical released into the environment result from a combination of factors including its innate toxicity and that of its biodegradation products as well as potential exposure to its harmful effects. Degradation of synthetic chemicals by microorganisms, biodegradation, is the most important breakdown process for chemicals deposited in the environment. This is due to the natural occurrence of diverse populations of microorganisms of high catabolic versatility and metabolic rate. In the best cases, biodegradation can result in the transformation of synthetic chemicals into innocuous compounds which are readily assimilated into the environment. In the

27

W. J. G. M. Peijnenburg and J. Damborský (eds.), Biodegradability Prediction, 27–40.
© 1996 *Kluwer Academic Publishers.*

28

worst case, an innocuous parent chemical could be transformed into a toxic intermediate. If a chemical is resistant to attack by microorganisms, it will generally persist in the environment, increasing its chances of adversely affecting humans, animals, and plants.

Microorganisms, naturally occurring in aquatic and terrestrial ecosystems, are capable of degrading a wide array of chemical structures. These microbes can degrade chemicals that are biochemically inert because of their ability to catalyze early steps in degradation, which other organisms cannot. In general, the initial degradation steps will lead to metabolites which can enter the common pathways of metabolism found in all living organisms. Aerobic microbes generally initiate degradation by utilizing enzymes that can incorporate one or two atoms of molecular oxygen into a substrate. Some specific reactions that aerobic bacteria use to initiate degradation are shown in Figure 1. After initiation, a number of additional reaction steps may be necessary to transform the parent compound to a metabolite that can enter a common channel of metabolism.

Figure 1. Examples of reactions that initiate degradation.

Monooxygenases:

$$CH_3\text{-}[CH_2]_6\text{-}CH_3 + O_2 + NADH + H^+ \xrightarrow{Fe^{2+}} CH_3\text{-}[CH_2]_6\text{-}CH_2OH + NAD^+ + H_2O$$

by Pseudomoma oleovorans

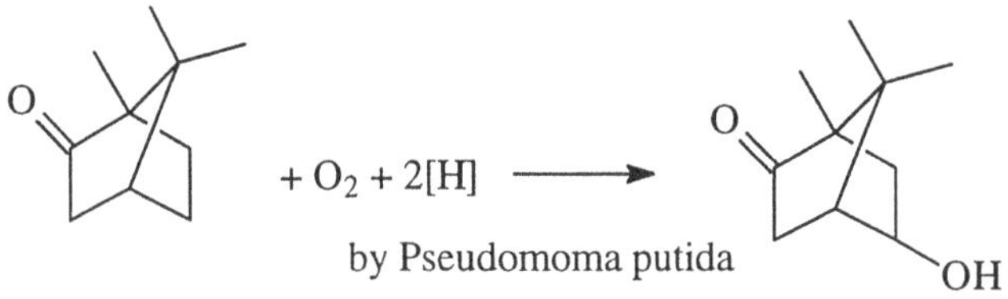

$+ O_2 + 2[H] \longrightarrow$ by Pseudomoma putida

Dihydroxylations:

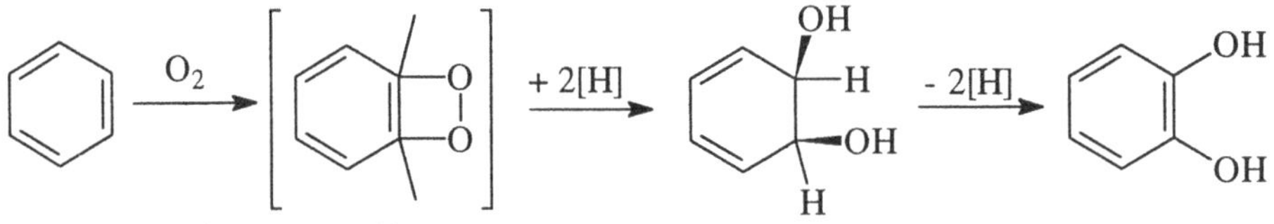

by Pseudomoma putida   cis-diol

Information on the metabolism of xenobiotic chemicals by soil microorganisms is usually obtained by isolating a pure culture capable of using the chemical of interest as a growth substrate. The metabolic pathway is then determined by identifying intermediate metabolites and characterizing the enzymes involved. This approach is justified because it appears that different microorganisms can use similar pathways in metabolizing the same chemical [1]. A wealth of knowledge on metabolic pathways of biodegradation has evolved from such studies. In general, microorganisms use an array of catabolic pathways that converge on a central metabolite. For example, Figure 2 shows that catechol is a central metabolite in the bacterial degradation of aromatic hydrocarbons.

Figure 2. The catechol pathway [2].

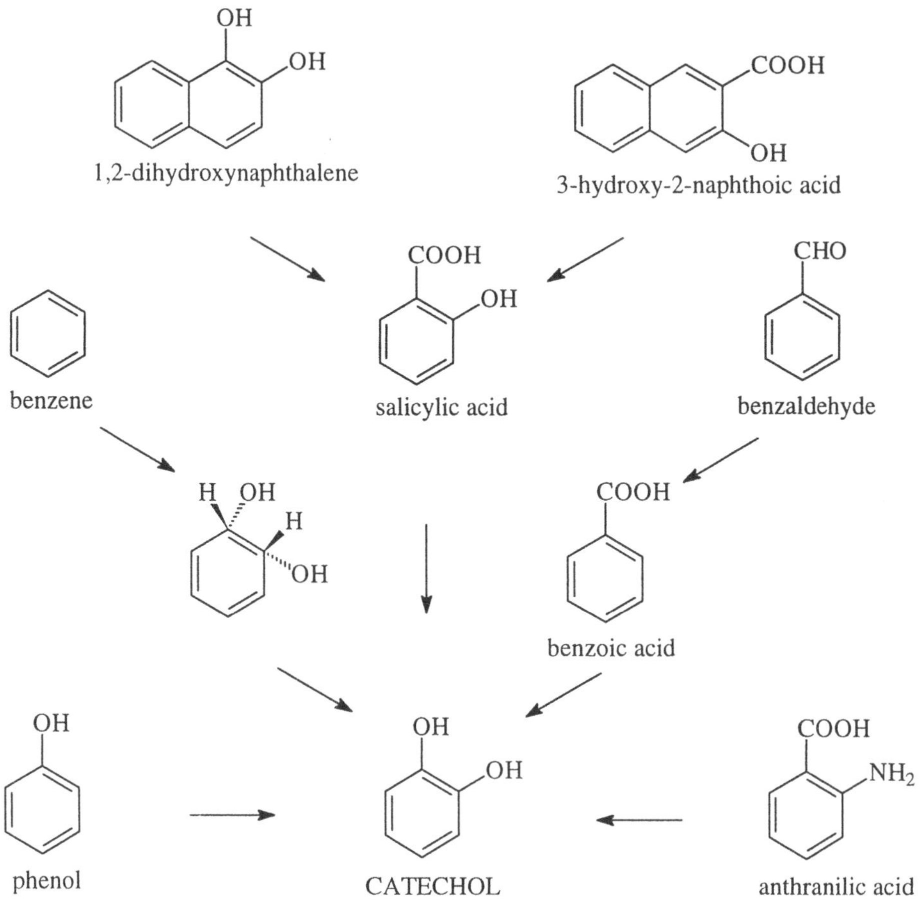

With the ever increasing use of chemicals by society, the number of existent and new chemicals which could find their way into the environment is tremendous. It is of utmost importance, therefore, to reduce adverse effects such as contamination, which could result from chemical release to the environment. Our objective, over the years, has been to develop a computer program for predicting the metabolic products formed from biodegradation of xenobiotic chemicals in the environment based on an expert evaluation of their chemical structure. By identifying what transformations can occur starting with the parent chemical, the possible end products of microbial degradation can be learned. This program will be referred hereafter as "META". The META program [3] provides a methodology that can quickly provide a user the structures of the metabolites expected from the microbial degradation of existing, new or proposed parent chemicals.

The META program is configured in such a way as to provide an interface with our MultiCASE program [4]. The MultiCASE program is an artificial intelligence program developed over the years in our laboratory to help uncover the complex relationships that may exist between the structure of chemicals and their biological activity. It has been successfully licensed to the pharmaceutical industry for drug design in the last five years. MultiCASE has been recently used to study chemical structure-biodegradation relationships of miscellaneous organic chemicals [5]. This lead to the identification of "activating" molecular fragments in molecules that underwent fast biodegradation.

In addition to identifying the most probable metabolites, we also evaluate their importance in the environmental fate of the parent chemical through an interface with MultiCASE. This is done by applying MultiCASE analysis to the META identified metabolites. Each rule in the dictionary also provides the user with the names of the microbial genera that have been reported to carry-out the transformation.

The resulting system can be used as a tool for the initial evaluation of the toxic potential of possible products or by-products of proposed manufacturing processes. Furthermore, the identification of structural features involved in biodegradation and microbial toxicity gives direct clues for modifying current or proposed fabrication methods. Ideal products or byproducts are compounds with features that are non-toxic, promote biodegradation and prevent microbial toxicity.

As an added benefit, the resulting predictive model can be used for developing a priority scheme to test molecules which are already being introduced into or are currently residing in the environment but have not received adequate toxicological evaluation. Compounds predicted to have a high toxicity risk should be assigned more resources for testing and treatment than those predicted to be harmless.

2. The MultiCASE Methodology

We have previously used the MultiCASE program to develop a preliminary structure-activity relationship (SAR) model for predicting the biodegradation potential of organic compounds under aerobic conditions [5]. A significant amount of work has been done by Howard [6] and coworkers in this field. They have used a set of 35 preselected structural features and successfully correlated the presence of these features in compounds with biodegradability data. A majority of our current data is from this work. However, unlike the model developed by Howard, the model generated using the MultiCASE method employs structural features that are automatically selected by the program from the information contained in a database of chemicals and their experimentally observed activities. As such, a dynamic model is created, limited only by the quality and availability of experimental data.

The MultiCASE program, just like its predecessor CASE [7], is a fully automated system that analyzes the biological activity of a given set of compounds and automatically identifies structural and physico-chemical descriptors believed to be responsible for the observed biological activity of the compounds (see flowchart in figure 3.).

Figure 3. The MultiCASE algorithm

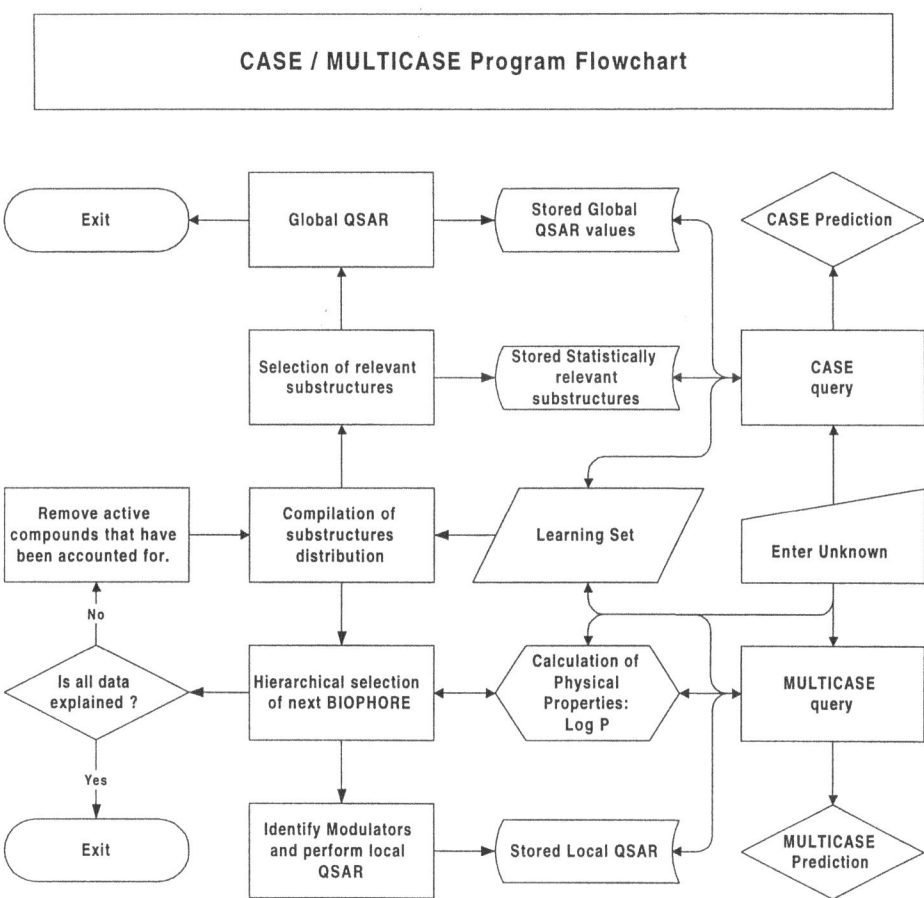

The molecular formulas of the compounds are entered graphically or by line notation schemes such as KLN [8] or SMILES [9]. Each compound to be analyzed is associated with a biological activity value. The values may be qualitative, i.e., active or inactive, or quantitative, e.g., IC50, percentage inhibition, LD50, etc. In the case of quantitative data, a cut-off point is selected which separates the actives from the inactives.

The resulting database is submitted as a learning set for MultiCASE analysis. MultiCASE takes each compound in the learning set and generates a set of fragments consisting of substructural and distance descriptors. Substructural descriptors are made of all possible fragments of two to ten linearly connected heavy atoms each together with attached hydrogens, hybridization information, and includes, if necessary, a single branch. Distance descriptors are made of all possible pairs of heteroatoms and heteroatoms with carbon atoms. A "distance" number is assigned to each pair depending on the smallest number of bonds between the atoms and the geometry of the shortest path between them. These substructural and distance fragments are labeled as active or inactive, depending on whether the parent molecule is active or not. The fragments are screened to determine which of them have a distribution that is markedly skewed towards either activity or inactivity. Essentially, fragments that have a binomial distribution (eq.1) of less than 15%, i.e., less than a 15% chance of being a random occurrence, are considered relevant.

$$\text{Probability } (x > x_0) = \sum_{x=x_0}^{n} \frac{n!}{x! \, (n-x)!} \, p^x \, (1-p)^{n-x} \tag{1}$$

where "Probability" is the probability that the observed distribution is due to chance, "n" is the total number of molecules containing a given molecular fragment, "x_0" is the number of these molecules that are active, and "p" is the overall percentage of active molecules.

The relevant fragments are kept as substructural descriptors to be used to create the predictive model. Among the fragments relevant to activity, the one that is found in the most actives and in the least inactives is selected. This fragment is called a *biophore*, a chemical functionality that is deemed responsible for activity. All the active compounds containing this biophore are removed from the database and set aside. The next biophore is selected from the remaining compounds and its related compounds are removed. This process is repeated until no more active compounds remain unexplained.

At this stage, the database has been separated into smaller data sets of congeneric compounds, each defined by a common substructural feature, i.e., a biophore. To a large degree, these biophores can be seen as structural alerts since they indicate a set of compounds showing a high percentage of active members. These subsets based on each of the biophores are analyzed separately. In addition to the pool of significant fragments generated earlier, physico-chemical property parameters such as lipophilicity (logP), [10] water solubility (ws) [11], graph indices (gi), and quantum mechanical indices (QM) [12,13], are calculated by the program. Using these as variables, a linear regression analysis is performed to yield a QSAR equation of the form: (eq.2)

$$\text{Activity} = a + \sum b_i(n_i F_i) + c \log P + d \log^2 P + e \, mwt + f \, ws + g \, gi + h(QM) \tag{2}$$

which explains the variation in experimental activities of compounds containing a given biophore. In this equation, a-h are regression coefficients, n_i is the number of times fragment i appears in a compound, F_i is equal to 1 if fragment i is present, otherwise it is 0, *log P* is the common logarithm of the partition coefficient between n-octanol and water, *mwt* is the molecular weight, and *ws* is the water solubility. In this equation, *log P* and *ws* are included to model the transport properties of the chemical. Also, *gi* is a topological graph index used to characterize the steric environment around the biophore. Finally, *QM* represents additional quantum mechanical parameters that include HOMO and LUMO coefficients, HOMO and LUMO energy gaps, ionization potentials and charge densities.

Each QSAR equation is valid only if the required biophore is present in the compound. The descriptors selected for the QSAR equation are called *modulators*. By themselves, modulators do not have the ability to cause activity. Rather, they serve to enhance or diminish the intrinsic activity imparted by the biophore. The collection of biophores and their corresponding QSAR equations are used to predict the potential activity of untested compounds. Their predictive ability is validated by applying the relevant equations to compounds with known activity that were withheld from the initial analysis.

The MultiCASE methodology has been applied successfully to the development of SAR models for a large number of toxicological endpoints. These endpoints include Rodent Carcinogenicity [14] , Salmonella Mutagenicity [15,16], Developmental Toxicity [17], etc. Our databases include all compounds evaluated by NIEHS under the National Toxicological Program (NTP). Statistically significant correlations have been obtained for each of them. Our validation tests typically consisted of leaving out fractions of the databases during analysis and evaluating statistical parameters such as the Chi^2 value, F-test values and concordance for the predictions of the activity of the molecules left out during the analysis.

3.   The META Methodology

We have developed a computer program called META for predicting the metabolic products formed from the biological transformation of chemicals, based on an expert evaluation of their molecular structure. The current program contains a dictionary of transformation rules which are applied to the parent chemical structures to form the metabolites. This dictionary/expert system provides an efficient methodology for predicting the biodegradation products of new chemicals since it is designed to recognize molecular fragments within molecules rather than complete structures. The META methodology is described as follows (figure 4).

The META program is a rule-based expert system designed to recognize substructural fragments in a given compound and replace the fragment with another fragment. Each rule is a pair of recognition and replacement fragments. Within the dictionary, each transformation rule is equivalent to one reaction step in a metabolic pathway. The recognition fragment consists of a string of two to ten linearly connected heavy atoms, similar to the substructural fragment used by the MultiCASE program. When a new chemical is submitted to the program, all possible substructural fragments are identified and compared to the recognition fragments contained in the dictionary. When a match is found the program executes that rule on the parent chemical to generate the resultant metabolite. Each metabolite can be submitted for further analysis, thus creating a metabolic tree of structures. A custom designed database management system keeps track of all the generated information allowing access to all the structures and transformations applied when requested.

34

Figure 4. The META methodology

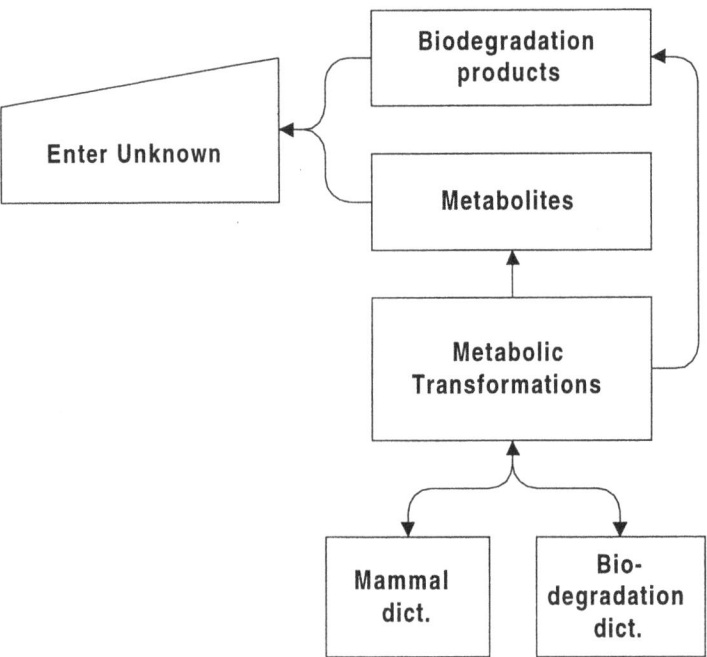

Aside from the recognition and replacement fragments, a rule also has a priority value, P, associated with it. Since in many instances more than one rule will apply to a given structure, priority values are used to decide which transformation rules are to be performed. Rule priorities run in the range from zero to nine, with zero being the highest priority. By convention, the transform with the highest priority P, and appropriate transforms with priorities P+1 and P+2 will operate on the chemical structures and be automatically shown on the screen. However, upon request, all transformations which hit the molecule will be shown. In the current version of the program, all the priorities were assigned by the authors, based on an evaluation of known experimental data. The criterion was to force the program to reproduce as closely as possible the observed order of importance of the metabolites resulting from the biodegradation of individual compounds. We are in the process of developing and evaluating different methodologies for the automated assignment of priorities.

3.1. DATABASE CONSTRUCTION

3.1.1. Aerobic Biodegradation

The data used in this part of the program was collected from all available literature sources. The construction of biodegradation databases is particularly difficult due to the wide variety of experimental conditions under which data is generated. The biodegradability of a compound is affected by several factors including microbe class or species, soil temperature, microbe population density, etc. The multitude of microbial species biodegrade chemicals in a number of different ways. Nonetheless, we hypothesized that these various avenues of biodegradation can be grouped into general classes of reaction mechanisms, such as monooxygenation, hydrolysis, etc.

In order to combine biodegradation results from various references, it is necessary to find compounds common among different sources to have bases for comparison. In the ideal case of having several common compounds, quantifiable relationships may be found between different sets of data such that quantitative data, e.g., rate of degradation, can be combined. However, when very few or no commonalities are found, a database of qualitative or binary data, i.e., compounds characterized as merely biodegradable or not biodegradable, can still be assembled.

Our current aerobic biodegradation data consists of 385 molecular structures. Given the diversity of the mechanisms involved and the wide variety of chemical classes covered, a much larger data set will be required to identify statistically significant parameters and formulate reasonably accurate models. We are currently working on reviewing and analyzing a more comprehensive database of over 500 compounds with experimental data on biodegradation rates and percent biodegradability taken from work by Pitter and Chudoba [18].

There also exists a number of genetically engineered bacteria, which were designed to target specific agents. Our current program does not specifically cover these bacteria. It is our intention however, to create appropriate modules to simulate biodegradation data related to genetically modified bacteria. However, since these genetic modifications only cover specialized metabolic paths, and generally do not suppress the commonly observed ones, we do not expect this exercise to be very difficult. The final outcome will eventually be the development of a number of dictionaries covering a variety of microbial metabolism variations. These dictionaries can then be combined to represent any particular environment.

It is recognized that the inability of microbes to biodegrade a particular compound may be due to several reasons. Primarily, the lack of suitable biodegradation biophores would prevent the microbial attack. Another reason is microbial toxicity. Even though a compound may have reasonable biodegradation biophores, it will not be biotransformed if the biodegrading microbes are killed in its presence. Hence, it is necessary to determine the biophores of microbial toxicity to gain a better understanding of why some compounds do not biodegrade.

3.1.2. Microbial Toxicity

Microbial toxicity data was collected from available literature. A CASE analysis has been done [19] using the results of the toxicity of a group of chemicals to four groups of bacteria by Blum and Speece [20]. A collection of approximately 150 compounds were tested in some or all of the assays involving aerobic heterotrophs, *Nitrosomonas*, methanogens, and the Microtox® test. The analysis resulted in the identification of significant biophores for each of the assays. However, the breadth of the coverage of the chemicals represented in the database was rather limited. We tested our microbial toxicity models with a test set of 4952 chemicals as a representative set of the "chemical universe." Warnings are generated for each of the compounds in the test set that contain substructures not found in the learning set. The results indicate that as much as 73% of the "chemical universe" is not represented in the current database. In contrast, our Salmonella Mutagenicity database, based on the National Toxicology Program's compilation, misses only about 14%. As such, these current microbial toxicity models will not be applicable to a large number of other chemical classes. This underscores the need for additional data.

3.1.3.   *Identification of Relevant Structural Parameters*

The MultiCASE analyses resulted in the identification of a wide variety of biophores believed to be responsible for the biodegradability of compounds. Furthermore, structural features responsible for the non-biodegradability of compounds due to microbial toxicity were also identified. Knowledge of these features gives a handle into their mechanisms of action. In fact, we have used the biophores identified in our analysis of chemical biodegradability in the development of our preliminary biodegradation dictionary for the META program. Our procedure and some preliminary results are presented in the next section.

3.2.   DICTIONARY CONSTRUCTION

The development of the dictionary was done by a stepwise process, based on the evaluation of the "rates" of biodegradation of known chemicals. First, we compiled a qualitative database that contained 385 organic chemicals, within which 172 molecules were known to biodegrade rapidly, and 213 molecules biodegrade slowly. MultiCASE was used to identify the relevant biophores. These biophores, when present in a molecule, indicate a high degree of biodegradability of the molecule. We interpreted these biophores as indicating substructures highly sensitive to attack by microorganisms, and therefore as descriptive of enzyme recognition sites. This set of biophores was therefore used as a starting point to uncover transformation rules from existing data on metabolic pathways. A transformation rule was generated around the structure of each of these biophores. For example, an ester group, a particularly good site for hydrolysis,

was identified by MultiCASE in 30 compounds, of which 22 (73%) biodegraded rapidly. Hence, the following rule was constructed:

```
Find:     CO -O  -CHn-
Replace: CO -OH   CHn- <3-OH>   (n=0-3)
```

where <3-OH> indicates the addition of a hydroxyl group to the atom at the 3-position. Also, note the removal of the bond between atom groups 2 and 3.
Another biophore shown below:

indicates the sensitivity of n-alkyl alcohols to oxidation. The outcome of the biotransformation is the corresponding carboxylic acid, presumably via the intermediacy of an aldehyde. Hence two transformation rules were written. One transformation corresponded to the oxidation of primary alcohols to aldehydes. The second rule was written to transform the resulting aldehyde to a carboxylic acid. These rules are illustrated as follows:

```
Find:      OH -CH2-CHn-
Replace: O   =CH -CHn- .        (n=1-2)

Find:      COH-CHn-
Replace: CO -CHn- <1-OH>        (n=1-2)
```

where COH is an aldehydic group and CO is a carbonyl group. These reactions have been documented as important steps in the bacterial degradation of aliphatic hydrocarbons [21]. For aromatic chemicals, transformation rules of hydroxylating aromatic rings were written corresponding to other biophores, and a series of transformation rules were also written to cover *ortho* and *meta* ring cleavage after hydroxylation.

The initial set of transformation rules was then tested with chemicals whose metabolites were known. This procedure led to the identification of other rules that needed to be incorporated into the dictionary, and helped us to adjust priorities and to evaluate the generality of the transformation rules. Finally, we went to the literature to identify biodegradation pathways for some other functionalities. For instance, we found biodegradation pathways for ketones [22], methoxy group on the aromatic rings [23], hydroxyl or carboxyl substituted pyridines [24], aryl methyl [25], amino acids [26], sulfates [27], imines [28], azos [29], etc.. Transformation rules reflecting these literature data were incorporated into our dictionary. Some of the resulting transforms are shown in Table 1.

Our current dictionary contains 85 transformation rules divided into 18 different activity classes, i.e., dioxygenase, esterase, amidase, deaminase, alcohol dehydrogenase, aldehyde dehydrogenase, beta-oxidation, sulfatase, dehydrase, monooxygenase, nitrile hydratase, nitriase, imine cleavage, azo reductase, halidohydrolase, muconate cycloisomerase and dienelactone hydrolase, maleylacetate reductase and ether hydrolase.

Table 1. Transformation rules for some functionalities

Reaction Type	Transformation Rule
Monooxygenation of ketones to form esters	F:*CHn-CO -CH2 R:*CHn-CO CH2 <2-O-3> (n=1-3)
Monooxygenation of methoxy group on aromatic ring	F: CH3-O -C = R: CH2-O -C = <1-OH >
Loss of ammonia from amino acids	F: NH2-CH -CH2- <2-CO > R: NH3 CH =CH - <2-CO >
Hydrolysis of alkyl sulfates	F: CHn-O -SO2 -OH R: CHn-OH SO2 -OH<3-OH>(n=1,2)
Cleavage of Imine	F: N =CH -CHn - R: NH2 COH -CHn -
Azo Reduction	F: CHn-N =N - R: CHn-NH2 NH2- (n=0-3)

* F means 'Find'; R means 'Replace with'

<a-X> means group X attached to atom nr. a.

<a-X-b> indicates that group X bridges atoms a and b.

In its current state, the rules are designed to be as general as possible. The definition of a recognition fragment can be further restricted though when new data to the contrary becomes available. Hence, the program will tend to predict more products than what may be experimentally observed. More work is in process to refine and extend the current rules to encompass a wider variety of compounds and aerobic microbial species. META, MultiCASE and CASETOX are trademarks of MULTICASE Inc. and the corresponding programs are available from MULTICASE Inc.[i]

4. References

1) Slater, J.H. and Lovath, D. (1984) Microbial Degradation of Organic Compounds, Ch. 15. Marcel Dekker, Inc. New York.

2) Chapman, P.J. (1972) Degradation of Synthetic Organic Molecules in the Biosphere. Natural, Pesticidal and Various other Man-Made Compounds. Natl. Acad. Sc. Washington D.C.

3) Klopman, G., Dimayuga, M., and Talafous, J. (1994) META [1]. A Program for the Evaluation of Metabolic Transformation of Chemicals. *J. Chem. Inf. Comput. Sci.* **34**:1320-1325.

4) Klopman, G. MultiCASE 1. (1992) A hierarchical computer automated structure evaluation program. *Quantitative Structure-Activity Relationships* **11**:176-184.

5) Klopman, G., Balthasar, D.M., and Rosenkranz, H.S. (1993) Application of the Computer-Automated Structure Evaluation (CASE) Program to the Study of Structure-Biodegradation Relationships of Miscellaneous Chemicals. *Environ. Toxicol. Chem.* **12**:231-240.

6) Howard, P.H., Boethling, R.S., Stiteler, W.M., Meylan, W.M., Hueber, A.E., Beauman, J.A., and Larosche, M.E. 1992) Predictive model for aerobic biodegradability developed from a file of evaluated biodegradation data. *Environ. Toxicol. Chem.* **11**:593-603.

[i] MULTICASE Inc., P.O.Box 22517, Cleveland, OH 44122, USA

7) Klopman, G. (1984) Artificial intelligence approach to structure-activity studies. Computer automated structure evaluation of biological activity of organic molecules. *J. Am. Chem. Soc.* **106**:7315-7320.

8) Klopman, G. and McGonigal, M. (1981) Computer simulation of physical-chemical properties of organic molecules. *J. Chem. Inf. Comput. Sci.* **21**:48-52.

9) Weininger, D. (1988) SMILES, a chemical language and information system. *J. Chem. Inf. Comput. Sci.* **28**:31.

10) Klopman, G., and Wang, S. (1991) A computer automated structure evaluation (CASE) approach to calculation of partition coefficient. *J. Comp. Chem.* **12**:1025-1032.

11) Klopman, G., Wang, S., and Balthasar, D.M. (1992) Estimation of aqueous solubility of organic molecules by the group contribution approach. Application to the study of biodegradation. *J. Chem. Inf. Comput. Sci.* **32**:474-482.

12) Houser, J.J., Klopman, G. (1988) A New Tool for the Rapid Estimation of Charge Distribution. *J. Comp. Chem.* **9**:893-904.

13) Houser, J.J., Klopman, G. (1992) The Rapid Estimation of Charge Distribution II. Heteroatoms. *J. Comp. Chem.* **13**:990-996.

14) Rosenkranz, H.S., and Klopman, G. (1990) Structural Basis of Carcinogenicity in Rodents of Genotoxicants and Non-Genotoxicants. *Mutat. Res.* **228**:105-124.

15) Klopman, G., and Rosenkranz, H.S. (1990) The Structural basis of the Mutagenicity of Chemicals in Salmonella typhimurium: The Gene-Tox Data base. *Mutat. Res.* **228**:1-50.

16) Rosenkranz, H.S., and Klopman, G. (1990) The Structural basis of the Mutagenicity of Chemicals in Salmonella typhimurium: The National Toxicology Program Data Base. *Mutat. Res.* **228**:51-80.

17) Takihi, N., Rosenkranz, H.S., Klopman, G., and Mattison, D.R. (1994) Structural Determinants of Developmental Toxicity. *Risk Anal.* **14**:649-657.

18) Pitter, P., and Chudoba, J. (1990) *Biodegradability of Organic Substances in the Aquatic Environment*, CRC press, Boca Raton.

19) Pangrekar, J., Klopman, G., and Rosenkranz, H.S. (1994) Expert-system comparison of structural determinants of chemical toxicity to environmental bacteria. *Environ. Toxicol. Chem.* **13**:979-1001.

20) Blum, D.J.W. and Speece, R.E. (1991) Quantitative structure-activity relationships for chemical toxicity to environmental bacteria. *Ecotoxicol. Environ. Saf.* **22**:198-224.

21) Britton, L.N. (1975) *Microbial degradation of organic compounds*. Chapter 5. Marcel Dekker, Inc. New York. (1984).Cripps, R.E. The microbial metabolism of acetophenone *Biochem. J.* **152**:233-241.

22) Cripps, R.E. (1975) The microbial metabolism of acetophenone. Biochem. J. 152:233-241.

23) Dagley, S. (1986) The biology of pseudomonas. *The Bacteria: a treatise on structure and function*, Sokatch, J.R. and Ornston L.N., eds. Academic Press, Inc. Orlando, **X**:415-416.

24) Sims, G.K., and O'loughlin, E.J. (1989) Degradation of pyridines in the environment. *Critical Review in Environmental Control* **19**:309.

25) Gibson, D.T. (1984) *Microbial degradation of organic compounds*, Marcel Dekker, Inc. New York. p107.

26) Phillips, A.T. (1986) The biology of pseudomonas. *The Bacteria: a treatise on structure and function*, Sokatch, J.R. and Ornston L.N., eds. Academic Press, Inc. Orlando, **X**:545.

27) Clarke, P.H. and Slater J.H. (1986) The biology of pseudomonas. *The Bacteria: a treatise on structure and function*, Sokatch, J.R. and Ornston L.N., eds. Academic Press, Inc. Orlando, **X**:88-94.

28) Sims, G.K., and O'loughlin, E.J. (1989) Degradation of pyridines in the environment. *Critical Review in Environmental Control* **19**:309.

29) Pitter, P., and Chudoba, J. (1990) Relationship between molecular structure and biological degradability in *Biodegradability of Organic Substances in the Aquatic Environment*, CRC press, Boca Raton, pp251-266.

APPLICATION OF ARTIFICIAL INTELLIGENCE IN BIODEGRADATION MODELLING

D. GAMBERGER, S. SEKUŠAK, Ž. MEDVEN, AND A. SABLJIĆ
Institute Ruder Bošković
P.O. Box 1016, HR-10001 Zagreb, REPUBLIC OF CROATIA

1. Abstract

The inductive machine learning method coupled with a set of experts' judgements and evaluated experimental biodegradation data were used to develop structural rules for ultimate biodegradation. Developed rules have been tested on BIODEG and MITI data. Both external validation tests have shown that the rules have solid predictive ability and performed better than other available methods. BIODEG and MITI data-bases were also used to develop new improved biodegradation rules. Two sets of developed biodegradation rules have very good classification ability, up to 90% for poorly biodegradable chemicals, and disclose structural features that either stimulate or hinder environmental biodegradation of organic chemicals.

2. Introduction

Biodegradation of commercial chemicals in natural waters, soil and sediments has attracted considerable attention from the environmental research community [1-13] since this is a sequence of processes which is not well understood. Furthermore, this is a key factor which determines the fate of chemicals in the environment [3-5]. Unfortunately, laboratory tests and in situ procedures are quite lengthy and measured data for ultimate biodegradability are available for a relatively small number of commercial chemicals [2, 4, 5]. Thus, particular interest and efforts have been focused in recent years on the development of qualitative and quantitative models for estimating ultimate biodegradability, under realistic environmental conditions, of various classes of commercial chemicals [2, 6-11]. Several detailed analysis and evaluations of published quantitative models [14-16] have shown that all such models are of limited assistance and that their predictive level is relatively low, generally below 70%. According to those extensive evaluation studies, quantitative models based on molecular fragments contributions approach seemed to be the best available today [14, 17]. However, their chemical or application domains are uncertain and consequently even the best available models should be applied with caution.

The major obstacle that precluded development of better and reliable biodegradation models in the past was the absence of standardized and consistent biodegradation data for various chemical classes. Recently, two data-bases of "high-quality"

W. J. G. M. Peijnenburg and J. Damborský (eds.), Biodegradability Prediction, 41–50.

biodegradation data became generally available. The first is the BIODEG data-base from the Syracuse Research Corporation with evaluated and standardized biodegradation data for about 300 diverse commercial chemicals [9, 17, 18]. The second data-base comprises the so-called MITI data [11], results of a single uniform biodegradation test for nearly 800 commercial chemicals. It is of interest to learn that those two data-bases do not overlap to a large extent, only about 120 chemicals being present in both data-bases. Thus, reliable biodegradation data are now available for nearly 1000 chemicals and the stage is set to develop as well as extensively validate a new generation of reliable biodegradation models. However, it must be also pointed out that reported biodegradation data are in contradiction for a small subset of overlapping chemicals.

The biodegradation data from both data-bases are discrete values, i.e. those chemicals are classified as biodegradable or nonbiodegradable. This type of measured biodegradation data is perfectly suited for a direct evaluation and development of qualitative models. On the other hand, this type of data is not well suited for evaluation and development of quantitative models since their output (results) are regularly continuous data which must be reinterpreted or reclassified first in order to make them compatible and comparable with BIODEG and MITI data.

Recently, an inductive machine learning method was applied to derive qualitative structure-biodegradation relationships [19-21] for commercial chemicals in the if-then-else form. Eleven simple structural descriptors were used to describe functional groups and structural segments which seem to be important for fast biodegradation [6, 7, 9]. The following simple logical rule, consisting of six structural features related by elementary logical tests was derived for fast biodegradation of commercial chemicals in the environment:

$$(j < 180)(k \neq 0) \text{ V } (e \neq f)[(g \neq k) \text{ V } (j < 135)(j > 95)] \tag{1}$$

where j is molecular weight, k is the number of C-O bonds, e indicates chemical that consists of only C, H, N, and O atoms, f indicates the presence of nitro group, g is the number of all rings present in a chemical. This simple rule combines three sets of structural requirements that are necessary for fast biodegradability:

(i)   *chemicals with one or more C-O bond and molecular weight below 180,*
(ii)   *chemicals built of C, H, N, and O atoms but without nitro group and their number of rings must be equal to or smaller than the number of C-O bonds*
(iii) *chemicals built of C, H, N, and O atoms but without a nitro group and their molecular weight must be in the range from 95 to 135.*

This rule has been evaluated on two sets of internally consistent biodegradation data [19, 20] and has shown a very good predictive performance.

It was very gratifying to find out that those sets of structural requirements can be rationalized through mechanistic arguments. The molecular weight cut-off point at 180 most probably indicates that it is very difficult or impossible for larger molecules to cross or penetrate through biological membranes and that permeability of biological membranes determines their rate of biodegradation. However, the molecular weight cut-of point at 180 seems to be somewhat low and may be in part also a reflection of the selection of training set chemicals. The structural rule clearly shows that the presence of C-O bonds is also very important (basic prerequisite) for fast biodegradation as well as their relationship to the number of rings in chemicals. Oxygenation is often a rate limiting step in biodegradation and it is a well known that mono- and di-oxygenases initiate cleavage of aromatic rings by attaching a hydroxyl group to them. Thus, the presence of C-O bonds, e.g. OH substituents on aromatic rings, facilitates oxygenation and degradation of many aromatic chemicals. Inclusion of nitro group requirement in the rule is also in line with experience that nitrochemicals usually have slow biodegradation.

The main objective of this study is to extensively evaluate the developed structural rule on two large data-bases of "high-quality" biodegradation data: BIODEG data-base [9, 17, 18] and MITI data [11]. The evaluation process has been performed on all data from individual data bases as well as on the chemical class by class basis. Results of this evaluation study indicate (point to) limitations of the developed structural rule and suggest areas where it can be improved. Finally an effort is made to apply the inductive machine learning method [19-21] to biodegradation data from the BIODEG and MITI data-bases in order to develop new, improved and extended structural rules for fast biodegradation under environmental conditions.

3. Methodology

The artificial intelligence method applied to develop rules for fast ultimate biodegradability is an example-based learning system [19, 22, 23]. The task of the developed example based learning system [19] is to generate a rule from a given set of examples. Each chemical (example) is described by a set of input variables (structural descriptors) and an output binary variable which has value 1 if the chemical has fast biodegradability or 0 if it has slow biodegradability. The number of input variables is determined by the number of structural descriptors.

There are two types of input variables: variables of quality (represented by a string) and variables of quantity (integer and floating-point variables). The elementary logical tests (ELT) that can be performed on variables of quality are: an input is equal to a constant string, an input is different from a constant string, an input is equal to another input of the same type, and an input is different from another input of the same type (examples: $var1 = var2$, $var3 \neq const1$). Both variables of quantity can be additionally compared by a magnitude (greater, smaller) and can be shifted by a constant when compared with the other variable of the same type (examples: $var4 = var5 + const2$, $var6 > var7 + const3$). It is possible to restrict the usage of some of the ELTs for all or groups of descriptors.

The problem of learning can be interpreted as a problem of finding a combinatorial rule that has arguments as variables and classes as function values. Thus, it is possible to generate a large number of different rules that will satisfy all learning examples. The assumption of learning systems is that the simplest rule has the greatest chance of having the highest average percent of correct predictions for test examples. At the beginning, the system determines the number of necessary basic logical elements that satisfy the whole learning set and after that the number of necessary basic logical elements for learning sets in which some of the chemicals were omitted. If the exclusion of a chemical reduces the number of basic logical elements in the minimal set, it is concluded that the example (chemical) is potentially incorrect or an outlier. When such chemical is removed from the learning set the process is repeated until a simple non-reducible solution is obtained.

The search for the minimal set of basic logical elements can be deterministic and heuristic. The first approach, based on an exhaustive search through all promising subsets of all basic logical elements, has the advantage that the solution generated is certainly the absolute minimum and that all equally good solutions will be found. The drawback is that the computation time grows exponentially with the number of chemicals and the number of all possible basic logical elements. The heuristic approach generates a solution much faster but it may not be the real minimum. In modelling biodegradation we have used the heuristic approach as a first choice. The deterministic generated solutions were used only to control how far the heuristic results are from the real minimum. It was noticed that the number of equally good minimal sets of basic logical elements generated by the deterministic algorithm can be used as a measure of the quality of the learning set. The great number of different possible solutions means either that a learning set is too small or that it includes several examples (chemicals) significantly different from the rest of the set, i.e. potentially incorrect examples.

The generated rule does not depend on the order of chemicals in the learning set or on duplicate examples (chemicals with identical set of input and output values) and the generated rule must satisfy all learning examples. This is impossible if there are contradictory learning examples. Before starting the actual rule generation process, the algorithm tests the learning set for duplicate examples and eliminates the copies. After that, the learning set is tested for contradictions (examples with all equal input values and different output value) and in such cases both examples are excluded.

As indicated in the introduction, biodegradation of commercial chemicals is a complex process which is not well understood and it is important to use all existing knowledge about it in order to generate the most reliable biodegradation rules. The example-based learning system developed in our laboratory and applied in this study is very flexible and it allows existing expert knowledge or rules of thumb on biodegradation processes to be built into the rule generation procedure. Expert knowledge can be formulated as the new logic elements or combinations of logic elements. Such expertise is not limited to structure-biodegradation relationships but it may be also related to the environmental factors that can be important for ultimate biodegradation under realistic environmental scenarios, i.e. the type of degradation process (hydrolysis, oxidation, dehalogenation, etc), the activity of the microorganisms involved in degradation, the biological or biochemical parameters like co-factors, and co-metabolism.

4. Evaluation of the Biodegradation Rule by the BIODEG Data-Base

Evaluation of the biodegradation rule (equation 1) developed by the application of example-based learning system [19, 20] was carried out firstly with the evaluated biodegradation data from the BIODEG data-base [18]. Evaluation results are presented in Table 1.

TABLE 1. A comparison of correctly predicted (equation 1) and evaluated (BIODEG data-base) biodegradation data. The BIODEG data-base includes biodegradation data of 293 chemicals, of which 185 have fast and 108 slow biodegradation. The percentage of correct predictions is shown in parenthesis. HYCAR represents all hydrocarbons and their halogenated derivatives

| Chemical Class | Number of Chemicals | | | |
| | Fast Biodegradation | | Slow Biodegradation | |
	predicted	BIODEG	predicted	BIODEG
All Chemicals	158 (85.4%)	185	76 (70.4%)	108
Alcohols	27 (100%)	27	0 (0%)	2
Amines	9 (90%)	10	1 (50%)	2
HYCAR	7 (58.3%)	12	22 (95.7%)	23
Pyridines	4 (80%)	5	4 (57.1)	7
Phenols	17 (89.5%)	19	6 (54.5%)	11
Anilines	3 (75%)	4	12 (85.7%)	14
Triazines		0	4 (80%)	5
Nitrobenzenes		0	5 (83.3%)	6
Acids	47 (94%)	50	13 (81.3%)	16
Esters	20 (64.5%)	31		0

Analysis of the evaluation results shows that developed biodegradation rule (equation 1) [19] performed quite satisfactorily, particularly for the chemicals that biodegrade fast. Predictability of chemicals with slow biodegradation was significantly lower. However, to put the predictive performance of the developed biodegradation rule in the right perspective it must be emphasized that it is based on only 5 structural characteristics and that results presented in Table 1 are results of an external evaluation and they give unbiased information on its predictive ability. For comparison, the internal evaluation of the BIODEG program [17], which uses 37 structural descriptors to estimate biodegradability and is classified as one of the best available methods today [16], shows only about 10% better performance than a simple logical rule (equation 1). Unfortunately, this result of internal evaluation gives more information on quality of the fitting procedure than on the predictive ability of the BIODEG program. External evaluation of the BIODEG program on a large set of MITI data has demonstrated that its predictive ability is only around 60%.

5. Evaluation of Biodegradation Rule by the MITI Data-Base

The second stage of the evaluation process was a comparison of predicted and measured MITI data [14] for 761 chemicals. The results are shown in Table 2.

Again, analysis of this external evaluation shows that the developed biodegradation rule (equation 1) [19] performs very well for the chemicals that biodegrade fast. Predictability of chemicals with slow biodegradation is significantly lower. Average predictive score is 75% and this result indicates that our developed logical rule seems to be the best available method for estimating biodegradability.

TABLE 2. A comparison of correctly predicted (equation 1) and measured MITI biodegradation data. The MITI data-base includes biodegradation data of 762 chemicals, of which 364 have fast and 398 slow biodegradation. The percentage of correct predictions is shown in parenthesis.

| Chemical Class | Number of Chemicals | | | |
| | Fast Biodegradation | | Slow Biodegradation | |
	predicted	measured	predicted	measured
All Chemicals	323 (88.7%)	364	248 (62.3%)	398
Acyclic	202 (97.6%)	207	45 (56.3%)	80
Aliphatic-monocyclic	6 (60.0%)	10	6 (54.5%)	11
Aliphatic-polycyclic	2(100%)	2	7 (77.8%)	9
Aromatic-monocyclic	70 (80.5%)	87	87 (55.8%)	156
Aromatic-polycyclic	13 (61.9%)	21	74 (77.1%)	96
Hetero-monocyclic	23 (79.3%)	29	14 (60.9%)	23
Hetero-polycyclic	7 (87.5%)	8	15 (65.2%)	23

6. Development of New and Improved Biodegradation Rules

Analysis of evaluation results has indicated that the developed rule (equation 1) is performing poorly for some classes or subgroups of chemicals. Clear examples are hydrocarbons, amines, acids and esters as well as aliphatic-monocyclic and aromatic-polycyclic chemicals in general. Thus, it seemed to be necessary to include several new

TABLE 3. List of 17 structural descriptors applied in developing biodegradation rules by the example-based learning system [19, 20]. Descriptors a, b, d, e, f, h, and i are binary type, c, g, k, l, m, n, o, p, and r are integers, while j is a real type descriptor.

Mark	Structural descriptors
a	Presence of heterocyclic nitrogen atom
b	Presence of ester, amide, or anhydride groups
c	Number of chlorine atoms
d	Bicyclic alkanes
e	Chemical composed only of carbon, hydrogen, nitrogen and oxygen atoms
f	Presence of nitro group
g	Number of rings
h	Presence of epoxy group
i	Primary alcohols and phenols
j	Molecular weight
k	Number of all C-O bonds
l	Number of tertiary amino groups
m	Number of quaternary carbon atoms
n	Number of C=C bonds
o	Number of aromatic amino groups
p	Number of acid groups
r	Number of ester groups

structural descriptors in the learning process in order to develop better biodegradation rules. A complete list of structural descriptors applied in developing the new biodegradation rules is shown in Table 3.

It was decided to develop biodegradation rules for each data-base separately since their biodegradation data are of different type as described in the Introduction.

The following set of rules, based on seven structural features related by elementary logical tests, have been derived for fast biodegradation of commercial chemicals in the environment from the BIODEG data-base.

$$(j <= 74) \tag{2a}$$
$$(p = 1)\,(g \neq 1) \tag{2b}$$
$$(m = 0)\,(k \neq 2)\,(k \neq 0)\,(e = 1) \tag{2c}$$
$$(m = 0)\,(j > 103)\,(e = 1)\,(k \neq 0) \tag{2d}$$
$$(j > 103)\,(k \neq 2)\,(k \neq 0)\,(g \neq 1)\,(g <= 3) \tag{2e}$$
$$(m = 0)\,(j > 103)\,(e = 1)\,(g <= 3)\,(o = 0) \tag{2f}$$
$$(m = 0)\,(j > 103)\,(e = 1)\,(g <= 3)\,(g \neq 1) \tag{2g}$$

This set of rules means that a chemical will biodegrade fast if any of the following conditions is satisfied:

(a) all chemicals with a molecular weight below 74

(b) all mono acids without rings

(c) all chemical built of C, H, N, and O atoms, having 1, 3 or more C-O bonds, but without quaternary carbons

(d) all chemical built of C, H, N, and O atoms, having at least one C-O bond and molecular weight above 103, but without quaternary carbons

(e) chemicals with a molecular weight above 103, having 1, 3 or more C-O bonds, and 2 or 3 rings

(f) chemicals with a molecular weight above 103, built of C, H, N, and O atoms, with 3 rings or less, but without either quaternary carbons or aromatic amino groups

(g) chemicals with a molecular weight above 103, built of C, H, N, and O atoms, with 2 or 3 rings, but without quaternary carbons

This set of rules based on only seven structural descriptors has correctly classified 80% of easily biodegradable chemicals and 90% of poorly biodegradable chemicals compiled in the BIODEG data-base. Our next research effort was to develop set of rules for fast biodegradation from the MITI data. The same set of 17 structural descriptors were used and the following set of biodegradation rules have been derived:

$$(m = 0)\,(k = 1)\,(g = 0) \tag{3a}$$
$$(b = 1)\,(n \neq 1)\,(e = 1) \tag{3b}$$
$$(b = 1)\,(m = 0)\,(g = 0) \tag{3c}$$
$$(b = 1)\,(m = 0)\,(e = 1)\,(g <= 1) \tag{3d}$$
$$(m = 0)\,(e = 1)\,(l = 0)\,(n \neq 1)\,(g = 0) \tag{3e}$$
$$(m = 0)\,(e = 1)\,(l = 0)\,(n \neq 1)\,(k \neq 0)\,(g <= 1) \tag{3f}$$

Again, in chemical terms those rules mean the chemical will biodegrade fast if any of the following conditions is satisfied:

(a) acyclic chemicals with one C-O bond, but without quaternary carbons

(b) esters, amides or anhydrides built of C, H, N, and O atoms, but without or with 2 C=C bonds

(c) acyclic esters, amides or anhydrides without quaternary carbons

(d) esters, amides or anhydrides built of C, H, N, and O atoms, having one ring or less, but without quaternary carbons

(e) acyclic chemicals built of C, H, N, and O atoms, but without either quaternary carbons or tertiary amino groups and without or with 2 C=C bonds

(f) chemicals built of C, H, N, and O atoms, acyclic or with 1 ring, with at least one C-O bond, but without either quaternary carbons or tertiary amino groups and without or with 2 C=C bonds.

The second set of rules, based also on seven structural descriptors, has correctly classified 76.6% of easily biodegradable chemicals and 89.2% of poorly biodegradable chemicals contained in the MITI data-base.

7. Discussion

Extensive evaluation of biodegradation rules (equation 1), developed by the application of an example-based learning system [19, 20], has clearly shown that developed rules as well as the new approach for developing structure-biodegradation relationships present a significant contribution to the area of risk assessment and priority settings. A set of simple rules based on only five structural descriptors has performed better by 12.5% on external validation than a highly complex QSAR model based on 37 structural descriptors and nonlinear function [17]. Thus, it seems that either the classical QSAR approach is not adequate for modelling qualitative biodegradation data or that the BIODEG program is far too overfitted (using many insignificant descriptors). As a result its estimates are questionable and the BIODEG program should be used with extreme caution.

Evaluation of biodegradation rules (equation 1) has also shown that these rules are performing poorly for some classes or subgroups of chemicals; i.e. hydrocarbons, amines, acids and esters as well as aliphatic-monocyclic and aromatic-polycyclic chemicals in general. New and improved biodegradation rules have been developed, with the extended list of structural descriptors, one for BIODEG data-base and one for MITI data. These results, although preliminary, present significant improvement over the first set of rules (equation 1) since classification is improved by at least 10%. Furthermore, new rules have up to 90% correct classification for poorly biodegradable chemicals and this result is very important for priority setting.

Finally, we have analyzed improved biodegradation rules (equations 2 and 3) in order to extract structural requirements which are important for either fast or slow biodegradation. First, a comparative analysis on structural descriptors of all three sets of biodegradation rules shows clearly that they agree to a large extent which structural features are important for biodegradation. Four out of five structural descriptors that are present in the previous set of rules (equation 1) are also found to be important for the new extended rules (equations 2 and 3). Furthermore, there is also a good agreement between two sets of rules developed in this study on structural features significant for either fast or slow biodegradation. To summarize, low molecular weight, presence of only C, H, N, and O atoms in a chemical, presence of C-O bonds, acyclic structures, as well as acid, ester, amide and anhydride functional groups seem to be stimulating features for biodegradation. On the other hand, the presence of rings, quaternary carbons, tertiary and aromatic amines, and a single C=C bond seem to retard the biodegradation process. It is fair to conclude that this result is to a large extent in line with existing expert knowledge on biodegradation, but it also yields new insight into structure-biodegradation relationships which should be evaluated further in the future.

8. Acknowledgment

This work was supported by grants 1-07-159 and 2-06-221 awarded by the Ministry of Science and Technology of the Republic of Croatia. This work was also supported by the U.S.-Croatian Science and Technology Joint Fund in cooperation with U.S. Department of Agriculture and Croatian Ministry of Science and Technology under Project Number JF-120.

9. References

1. Efroymson, R.A. and Alexander, M. (1994) Role of partitioning in biodegradation of phenanthrene dissolved in nonaqueous-phase liquids, *Environ. Sci. Technol.* **28**, 1172-1179.
2. Peijnenburg, W.J.G.M. (1991) The use of quantitative structure-activity relationship for predicting rates of environmental hydrolysis processes, *Pure Appl. Chem.* **63**, 1667-1676.
3. Klečka, G.M. (1985) Biodegradation, in W.B. Neely and G.E. Blau (eds.), *Environmental Exposure from Chemicals, Vol. I*, CRC Press, Boca Raton, pp. 109-155.
4. Organization for Economic Cooperation and Development (1989). *Guidelines for testing chemicals - Section 3: Degradation*, Paris, France.
5. Boethling, R.S., Gregg, B., Frederick, R., Gabel, N.W., Campbell, S.E., and Sabljić, A. (1989) Expert systems survey on biodegradation of xenobiotic chemicals. *Ecotox. Environ. Safety* **18**, 252-267.
6. Boethling, R.S. and Sabljić, A. (1989) Screening-level model for aerobic biodegradability based on a survey of expert knowledge., *Environ. Sci. Technol.* **23**, 672-679.
7. Niemi, G.J., Veith, G.D., Regal, R.R., and Vaishnav, D.D. (1987) Structural features associated with degradable and persistent chemicals, *Environ. Toxicol. Chem.* **6**, 515-527.
8. Desai, M.D., Govind, R., and Tabak, H.H. (1990) Development of quantitative structure-activity relationships for predicting biodegradation kinetics, *Environ. Toxicol. Chem.* **9**, 473-477.
9. Boethling, R.S., Howard, P.H., Meylan, W.M., Stiteler, W.M., Beauman, J.A., and Tirado, N. (1994) Group contribution method for predicting probability and rate of aerobic biodegradation, *Environ. Sci. Technol.* **28**, 459-465.
10. Klopman, G., Balthasar, D.M., and Rosenkranz, H.S. (1993) Application of the computer-automated structure evaluation (CASE) program to the study of structure-biodegradation relationships of miscellaneous chemicals, *Environ. Toxicol. Chem.* **12**, 231-241.

50

11. Takatsuki, M., Takayanagi, Y., and Kitano, M. (1995) An attempt to SAR of biodegradation, in W.J.G.M. Peijnenburg and W. Karcher (eds.), *Proceedings of the Workshop "Quantitative Structure Activity Relationships for Biodegradation"*, National Institute of Public Health and Environmental Protection (RIVM), Bilthoven, The Netherlands, pp. 67-103.

12. Castro, C.E. (1993) Biodehalogenation: the kinetics and rates of the microbial cleavage of carbon-halogen bonds, *Environ. Toxicol. Chem.* **12**, 1609-1618.

13. Niemela, J.R. (1994) Validation of the BIODEG probability program, *TemaNord.* **589**, 153-156.

14. Degner, P., Müller, M., Nendza, M., and Klein, W. (1993) *Structure-activity relationships for biodegradation*, OECD Environment monographs No. 68.

15. Hermens, J., Balaz, S., Damborský, J., Karcher, W., Müller, M., Peijnenburg, W., Sabljić, A., and Sjöström, M. (1995) Assessment of QSARs for predicting fate and effects of chemicals in the environment: an international european project, *SAR QSAR Environ. Res.* **3**, 223-236.

16. Langenberg, J.H., Peijnenburg, W.J.G.M., and Rorije, E. (1996) On the usefulnessand reliability of existing QSBRs for risk assessment and priority setting, *SAR QSAR Environ. Res.* **5**, 1-16.

17. Syracuse Research Corporation (1992) *Biodegradation probability program (BIODEG), Version 3*, Syracuse, NY.

18. Syracuse Research Corporation (1995) *Environmental fate data base,* Syracuse, NY.

19. Gamberger, D., Sekušak, S., and Sabljić, A. (1993) Modelling biodegradation by an example based learning system, *Informatica* **17**, 157-166.

20. Gamberger, D., Sekušak, S., and Sabljić, A. (1995) Application of expert judgement to derive structure-biodegradation relationships, in W.J.G.M. Peijnenburg and W. Karcher (eds.), *Proceedings of the Workshop "Quantitative Structure Activity Relationships for Biodegradation"*, National Institute of Public Health and Environmental Protection (RIVM), Bilthoven, The Netherlands, pp. 61-66.

21. Gamberger, D., Horvatić, D., Sekušak, S., and Sabljić, A. (1996) Application of expert judgement to derive structure-biodegradation relationships, *Environ. Sci. Poll. Res.* **3**, in press.

22. Gamberger, D. (1995) A minimization approach to proportional inductive learning, in N. Lovrac and S. Wrobel (eds.), *Proceeding of 8th European Conference on Machine Learning, Springer Lecture Notes in Artificial Intelligence*, pp. 151-160.

23. Kompare, B., Bratko, I., Steinman, F., and Džeroski, S. (1994) Using machine learning techniques in the construction of models, *Ecological Modelling* **75/76**, 617-628.

POLYCHLORINATED DIBENZO-*p*-DIOXINS IN ANAEROBIC SOILS AND SEDIMENTS

A Quest For Dechlorination Pattern-Microbial Community Relationships

P. ADRIAENS, A.L. BARKOVSKII and M. LYNAM
Environmental and Water Resources Engineering, Department of Civil and Environmental Engineering, The University of Michigan, Ann Arbor, MI 48109-2125 USA

and

J. DAMBORSKÝ and M. KUTÝ
Department of Environmental Studies, Masaryck University, Kotlarska 2, 611 37 Brno, CZECH REPUBLIC

1. Abstract

Significant differences have been observed in the 2,3,7,8-substituted residue patterns of polychlorinated dibenzo-*p*-dioxins (PCDD) in freshwater, estuarine and marine sediments. Whereas these patterns can, to some degree, be explained by source identification, PCDD at environmental concentrations were recently found to be dechlorinated via microbial and chemical processes. Both *peri*- (1,4,6,9-substituted chlorines) and lateral (2,3,7,8-substituted chlorines) dechlorination patterns, as well as differences in extent of dechlorination were found to be correlated to specific abiotic and biotic catalytic activities. Qualitative relationships were based on isomer-specific analysis and the appearance of selective congeners under different conditions. The relevance of these processes to sediment biogeochemistry indicates that microbial dechlorination contributes significantly to the natural weathering of these types of pollutants. Whereas the lack of knowledge on the catalytic nature of the dechlorination reaction precludes the establishment of QSARs, characterization of microbial activities in combination with geochemical indicators may eventually present a means to describe the potential for microbial PCDD dechlorination in a given sediment environment and allow for a scientifically justified interpretation of patterns observed.

2. Dioxins in the Environment

Polychlorinated dibenzo-*p* -dioxins (PCDD) and dibenzofurans (PCDF) are among the most hazardous environmental pollutants due to their estrogenicity and suspected

51

W. J. G. M. Peijnenburg and J. Damborský (eds.), Biodegradability Prediction, 51–64.
© 1996 *Kluwer Academic Publishers.*

carcinogenicity, as well as their potential for bioaccumulation and persistence in various ecosystems. Depending on the chlorination pattern, up to 75 dioxin congeners and 135 dibenzofuran congeners can theoretically be formed (Table 1). The laterally substituted isomers (chlorines in 2,3,7,8-positions) are considered to exhibit the highest toxic activity, and their concentrations in sediments are on the order of pg.kg^{-1} to ng.kg^{-1} for 2,3,7,8-tetraCDD/F alone, and ng.kg^{-1} to µg.kg^{-1} for all 2,3,7,8-substituted PCDD/F (excluding octaCDD/F). These compounds were never intentionally manufactured, but are formed (i) during chlorophenol and chlorinated herbicide (e.g. 2,4-D and 2,4,5-T) synthesis, (ii) as toxic components of wood treating wastes derived from the use of pentachlorophenol, (iii) as byproducts during bleach processes in the pulp- and paper industries, and (iv) during incineration processes[17,39,34].

The environmental burden of waterways and sediments with chlorinated dioxins and furans, as well as the PCDD/PCDF congener distribution is dependent on the point sources of discharge and the process which generated them. Generally, the concentrations of PCDD/PCDF in sediments are two to three orders of magnitude lower than those of polychlorinated biphenyls (PCBs). In studies on the fate of PCDD/F in their movement from source to sink, it has been observed that the corresponding homologue profiles of PCDD/F from source and sink are different. Overall, environmental PCDD/F 'fingerprints' are characteristic for the process that generated them. Thus, incineration source profiles have been shown to be more uniform (i.e. an even distribution over the whole range of homologues) than the profiles generated from chlorophenol or 2,4,5-T synthesis (highly enriched in 2,3,7,8-substituted isomers). Sink profiles, on the other hand, have been found to be dominated by either the higher chlorinated PCDD/F, especially octaCDD and heptaCDF or by 2,3,7,8-substituted isomers in general. It was found that mainly wet deposition processes scavenge particle-bound highly chlorinated PCDD/F in the atmosphere resulting in the enrichment of these congeners in the terrestrial sinks [21].

Despite the information on 2,3,7,8-substituted dioxin residue patterns from known sources and those from freshwater and estuarine sediments, source-sink correlations as analyzed by chemometric and polytopic vector analysis of 2,3,7,8-substituted PCDD/F residues in surficial sediments have repeatedly failed to establish a link to industrial point sources [45,43]. An example of a three-dimensional scores PCA (Principal Component Analysis) plot for 2,3,7,8-substituted dioxin residue patterns in international industrial waterways is shown in Figure 1 (after Wenning et al., 1992). Each plane represents a principal component, containing a standardized set of variables that best describes the relationships among the samples and the variance contained in the original data. Thus, sediment samples are clustered according to similarities and differences among isomeric patterns. A principal components loadings plot of individual isomers provides information on which congeners best explain the differences observed between groups of samples (Figure 1, B). This shows that octa- and heptaCDD/F, and particularly the 2,3,7,8-substituted isomers, are responsible for the differences between groups, as they are located furthest from the origin (all other isomers are clustered around the origin). It has been suggested that either a multitude of sources [30,44] or environmental transformation reactions resulting in 'weathering' [25] may be responsible for the limited successes in establishing source/sink correlations for dioxin residues in sediments.

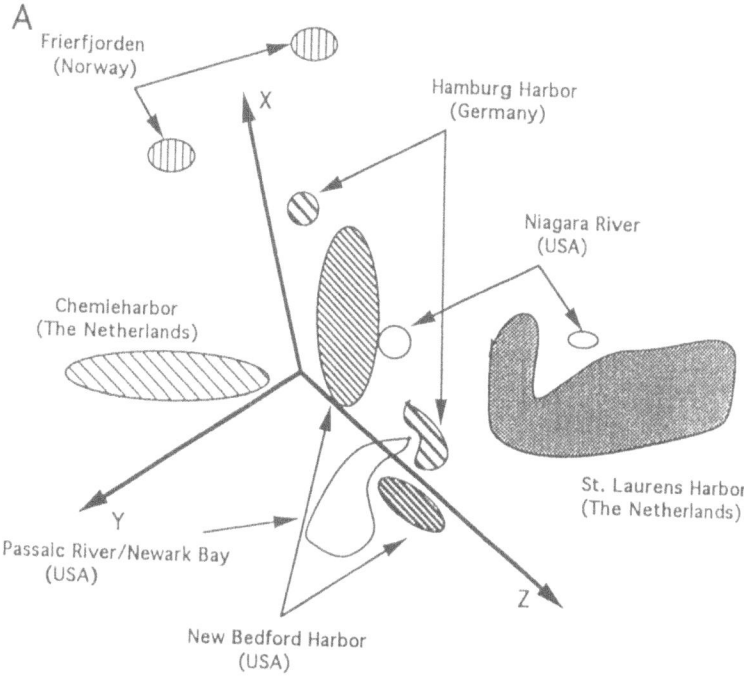

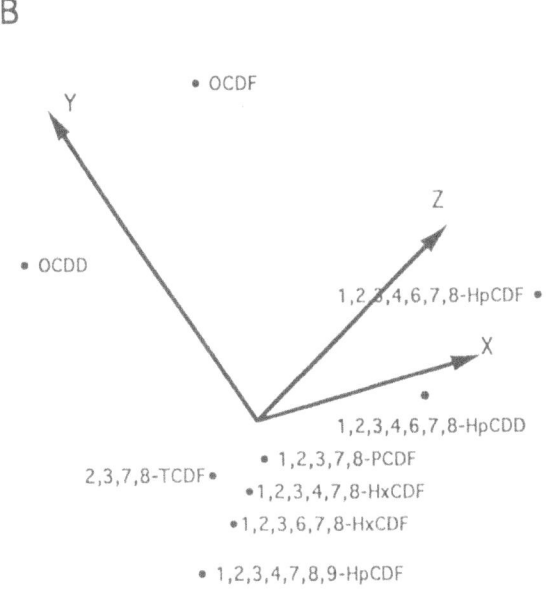

Figure 1.   Three Dimensional Scores Plot from Principal Component Analysis (PCA)
Indicating Dioxin Pattern Clusters for Industrial Waterways (A), Loadings Plot of Normalized
2,3,7,8-Substituted PCDD/F (B) (redrawn from Wenning et al., 1992).

TABLE 1.   Structure and number of isomers within each dioxin/furan homologue group

Level of Chlorination	Total Isomers	Laterally Substituted Isomers*	Total Isomers	Laterally Substituted Isomers
Mono-	2	-	4	-
Di-	10	-	16	-
Tri-	14	-	28	-
Tetra-	22	1	38	1
Penta-	14	1	28	2
Hexa-	10	3	16	4
Hepta-	2	1	4	2
Octa-	1	1	1	1

* Chlorines at positions 2,3,7, and 8

2.1. BIOGEOCHEMISTRY OF ANOXIC SEDIMENTS

Microbial diversity and metabolic activity in estuarine and marine environments is affected by salinity gradients, high sulfate concentrations, and competition for energy sources. Differences in the composition and content of organic material between freshwater and marine (or estuarine) sediments likely affect bacterial activities. For example, trends have been observed in the bromine to organic carbon ratio of sedimentary organic material between freshwater and coastal areas, which was attributed to different source materials. The effect of ionic strength on microbial populations is dependent on the osmoregulatory mechanisms evolved in response to salt stress. Whereas halotolerant bacteria do not require the high NaCl concentrations for growth but can grow under saline conditions, halophiles require NaCl for growth [35]. Salinity gradients between 5-20 ‰ reportedly exerted a negative effect on freshwater aerobic heterotrophic [18, 38,36], methanotrophic [22], and anaerobic [35,32] respiratory activities.

Because of the high sulfate concentrations in saline environments (20-30 mM), sulfate-reduction is suspected to be the dominant process in carbon metabolism in marine sediments [40, 27] although fermentation [35], denitrification [13,32], iron reduction [29, 28], and methanogenesis [22, 35] have been demonstrated. Methanogenesis in marine environments is not linked to hydrogen consumption, but occurs in environments where methylamines are used as the non-competitive substrates [35]. At salt concentrations above 15‰, sulfate-reduction and methanogenesis were negatively impacted, resulting in a buildup of H_2 and acetate, due to incomplete substrate oxidation.

In addition to aquatic (pore-water) nutrient composition, the establishment of terminal electron processes in sediment depth profiles can be expected to be strongly correlated with sedimentation rates [26]. Indeed, spatial and temporal variations in microbial processes have been observed in terrestrial, estuarine, and coastal aquatic sediments: aerobic and denitrifying activity is confined to the top few centimeters of sediments, sulfate-reduction has been observed over 50-60 cm of sediment thickness, which is underlain by a zone of methanogenic activity [40,15]. Thus, sulfate-reduction is expected to be the dominant process responsible for organic matter oxidation to carbon dioxide in estuarine and coastal sediments, unless high sedimentation ratios prevail and methanogenic conditions develop in the deeper layers [40]. Sulfate-reducing processes significantly impact porewater and sediment chemistry, including a decrease in pH with a concurrent rise in alkalinity and sulfides, carbonate precipitation, ammonia production, and reduction of iron(III) hydroxide minerals by sulfides.

2.2. REDUCTIVE DECHLORINATION REACTIONS IN ANOXIC SEDIMENTS

River and lake sediments are often anaerobic and very reduced, which renders them ideal habitats for anaerobic microbial populations. Moreover, functional groups from humic acids (e.g. phenols, hydroquinones), heavy metals (e.g. Zn), biogenic factors (e.g. vitamin B_{12}), and reduced mineral surfaces constitute potentially significant electron shuttles in reduced environments. Dioxins, as carbonaceous organic material, are subject to the same environmental variables which affect the biogeochemical carbon cycle, including oxidation and reduction reactions as well as burial with particulate or dissolved organic matter during sedimentation processes. They would thus be expected to undergo reactions catalyzed under the different terminal electron accepting processes as they develop during sedimentation.

Reductive dechlorination reactions result from electron transfer between a reduced organic or inorganic electron donor and the generally highly oxidized polychlorinated target molecule, such as PCDD. Whereas it is argued that this process is predominantly thermodynamically driven, the dechlorination kinetics, as well as substituent effects and extent of dechlorination, are highly dependent on environmental characteristics, and the dominant terminal electron accepting processes [1,2]. For example, six characteristic and recurring microbial dechlorination patterns have been identified for polychlorinated biphenyls (PCBs) in contaminated sediments around the world, yet very limited information is available on the basis for these differences [11]. The possibility that PCB dechlorination may not be a fortuitous process is gaining support based on (i) inducibility of PCB dechlorinating activity using specific electron donors [11], (ii) inducibility of specific dechlorination patterns using selected PCB congeners [10], (iii) the dependence of dechlorination patterns on microbial respiratory activity, and (iv) genetic differences between *meta*- and *para*-dechlorinating mixed populations [16,24].

Evidence for the dependence of PCB dechlorination patterns on dominant electron accepting processes was demonstrated recently. The dominant *meta/para* - dechlorination pattern of higher chlorinated congeners in Aroclors 1242 and 1254 (pattern H) observed in marine and estuarine sediments was distinctly different from that observed in freshwater sediments [14]. Ofjord et al. [33] demonstrated that cultures derived from marine sediments, and incubated under high sulfate conditions, were responsible for the *meta/para* dechlorination pattern when grown on chitin, a ubiquitous carbon source in marine environments. This observation underscores the ecological impact on ionic strength of seawater and the presence of approximately 27 mM of sulfate may have on the types of biodegradative communities involved, and indicates that reductive dechlorination rates and pathways in freshwater sediments may not reflect those prevailing in estuarine or marine environments. The above discussion furthermore indicates a need to include biogeochemical information on microbial activity in the development of structure-activity relationships, particularly in those cases where no mechanistic information is available on enzyme activity or specificity.

3. Microbial and Abiotic Dechlorination of Dioxins

3.1. DECHLORINATION PATTERNS

Environmentally-relevant concentrations of PCDD/F were recently demonstrated to undergo dechlorination in reduced freshwater sediments under methanogenic conditions [4, 3]. Estuarine[6,8] and freshwater [4, 12] sediment-derived microorganisms, humic acid constituents [7], biogenic macromolecules and heavy metals [1, 2] were found to catalyze dechlorinating activities. The specificity of dechlorination pathways was very dependent on the microbial amendment applied (pasteurization, model humic acid components) and the contamination history and source of the sediments used. Sediment characteristics are presented in Table 2. All inocula samples were incubated under conditions promoting methanogenesis.

Interestingly, methane production was consistently depressed upon amendment with freshly spiked PCDD in both freshwater and estuarine inoculum sources, except for in freshwater sediments (upper Hudson River). Whereas historically-contaminated communities maintained methanogenic activity longer, methane production eventually ceased as well [6]. These observations are consistent with reports on dechlorination of aryl halides under methanogenic conditions, where it is hypothesized that chlorinated aromatic compounds can serve as terminal electron acceptors, and thus scavenge electrons from bicarbonate reduction (reviewed in [1]). The on-going methanogenesis in sediments is speculated to be due to acetoclastic methanogens, which use the methyl group from acetic acid to form methane. Methanogenic activity ceased immediatelly upon pasteurization.

Four distinctive pathways could be discerned based on isomer-specific analysis: (i) *peri*-dechlorination of 2,3,7,8-substitued congeners resulting in the transient production of 2,3,7,8-TCDD, (ii) a relatively slower mixed *peri*-/lateral-dechlorination pathway for non-2,3,7,8-substituted congeners, (iii) lateral dechlorination of historically-partitioned 2,3,7,8-TCDD, and (iv) predominant peri-dechlorination in pasteurized cells (Figure 2).

TABLE 2.   Sources and contamination history of PCDD-dechlorinating inocula

Source	Classification	Contaminant History*	PCDD Spike Received	Reference
Hudson River (H7 Site)	Freshwater	PCBs	1,2,3,4,6,7,8-HpCDD 1,2,3,4,7,8-HxCDD	Adriaens and Grbic'-Galic' (1994); Adriaens et al. (1995)
ACW Aquifer (Pensacola)	Freshwater	Creosote	1,2,3,4,6,7,8-HpCDD 1,2,3,4,7,8-HxCDD	*ibid.*
Cherokee Pond (Athens, GA)	Freshwater	Chlorophenols	1,2,3,4,6,7,8-HpCDD	Adriaens and Grbic'-Galic' (1994)
Lake Ketelmeer (Netherlands)	Freshwater	Hexachloro benzene	1,2,3,4-TeCDD	Beurskens et al. (1995)
Lower Passaic River	Estuarine	PCDD, PCB PAH	1,2,3,4,6,7,8,9OCDD	Barkovskii and Adriaens (1996); Adriaens and Barkovskii, 1995)
Lower Passaic River	Estuarine	PCDD, PCB PAH	Historical Contamination	Barkovskii and Adriaens (1995)

* PCB, Polychlorinated biphenyls; PCDD, polychlorinated dibenzo-p-dioxins; PAH, polycyclic aromatic hydrocarbons.

Since *peri*-dechlorination of heptaCDD was also found to be important in freshwater sediments [4], these results may indicate that 2,3,7,8-residue patterns found in historically contaminated sediments may be the cumulative result of source patterns, and microbiologically-mediated natural attenuation processes. Further resolution of the differences observed between treatments were based on the extent of dechlorination, the relative predominance of 2,3,7,8-substituted to non-2,3,7,8-substituted congeners, as well as the appearance of selected individual congeners.

TABLE 3.   Level of octaCDD dechlorination during different microcosm incubation techniques

Microcosm Type	Extent of Dechlorination	Dominant Pattern	Most Abundant Congener Group	2,3,7,8-TCDD Formed[2]
Sediments	tetra[1]	*peri*	tetra	ND
Eluted Cell Amendments				
None	mono	lateral >*peri*	tetra, tri	+++
Pasteurized	tri	lateral	tri	-
Catechol	mono	lateral >> *peri*	tetra	+
Resorcinol	mono	lateral >> *peri*	tetra	++
3,4-Dihydroxy-benzoate	mono	*peri* >> lateral	hepta	+++
Abiotic Dechlorination				
Catechol	tetra	lateral >> *peri*	tetra	+
Resorcinol	tetra	lateral >> *peri*	tetra	++
3,4-Dihydroxy-benzoate	tetra	*peri* >> lateral	tetra	+++
Biomimetic				
Ti- citrate + vitamin B_{12}	tetra	ND[3]	tetra	+
w/o vitamin B_{12}	tetra	ND	penta	-
Zero-valent Zn				
alkaline (pH 9)	tetra	ND	hepta	ND
neutral (NH_4Cl)	tetra	ND	tetra	ND
neutral (H_2O)	hexa	ND	hepta	ND

[1] Not analyzed for mono- to trichlorinated dioxins

[2] Formation denoted as fraction of total TCDD formed: +++; > 25%, ++; >10%, +; < 10%.

[3] ND, Not Determined.

The dechlorination of octa- and pentachlorinated dioxins and dibenzofurans was investigated using organic and inorganic electron shuttles relevant to anaerobic sediments. OctaCDD was dechlorinated to hexaCDD in the presence of resorcinol, catechol, and 3,4-dichlorobenzoate, resulting in an increase of 2,3,7,8-substituted hepta- and hexaCDD. Whereas catechol and resorcinol showed a dechlorination pattern similar to that of organic compound-amended cells, dihydroxybenzoate incubations exhibited a significantly different pattern. Vitamin B_{12} mediated the dechlorination of octaCDD to tetraCDD, octaCDF to hexaCDF, and 1,2,3,7,8-pentaCDD/F to at least two tetraCDD, including 2,3,7,8-TCDD. Zerovalent zinc stoichiometrically dechlorinated octaCDD to hexa- and pentaCDD under basic and neutral conditions, respectively. Thus, abiotic dechlorination reactions may contribute significantly to the fate of PCDD/F in reduced environments, and may be distinguished from biological processes based on isomer-specific 'fingerprinting' of the dechlorination pattern (Table 3).

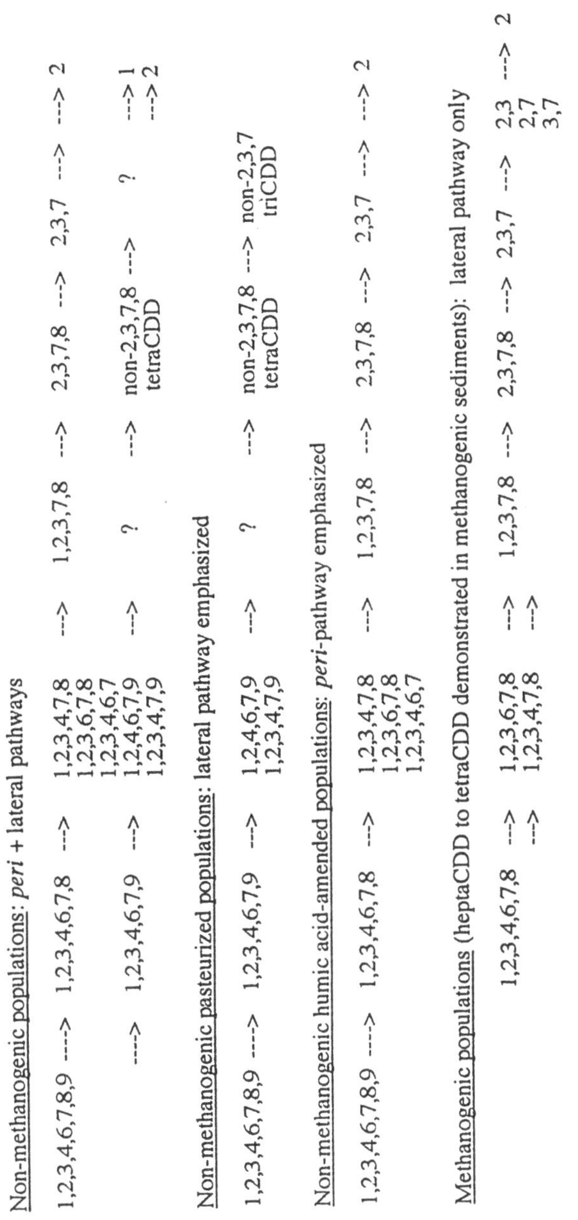

Non-methanogenic populations: *peri* + lateral pathways

1,2,3,4,6,7,8,9 ---> 1,2,3,4,6,7,8 ---> 1,2,3,4,7,8 ---> 1,2,3,7,8 ---> 2,3,7,8 ---> 2,3,7 ---> ---> 2
                          1,2,3,6,7,8
                          1,2,3,4,6,7
---> 1,2,3,4,6,7,9 ---> 1,2,4,6,7,9 ---> ? ---> non-2,3,7,8 ---> ? ---> ---> 1
                  1,2,3,4,7,9   tetraCDD   ---> ---> 2

Non-methanogenic pasteurized populations: lateral pathway emphasized

1,2,3,4,6,7,8,9 ---> 1,2,3,4,6,7,9 ---> 1,2,4,6,7,9 ---> ? ---> non-2,3,7,8 ---> non-2,3,7
                               1,2,3,4,7,9   tetraCDD   triCDD

Non-methanogenic humic acid-amended populations: *peri*-pathway emphasized

1,2,3,4,6,7,8,9 ---> 1,2,3,4,6,7,8 ---> 1,2,3,4,7,8 ---> 2,3,7,8 ---> 2,3,7 ---> ---> 2
                         1,2,3,6,7,8
                         1,2,3,4,6,7

Methanogenic populations (heptaCDD to tetraCDD demonstrated in methanogenic sediments): lateral pathway only

1,2,3,4,6,7,8 ---> 1,2,3,6,7,8 ---> 2,3,7,8 ---> 2,3,7 ---> 2,3 ---> 2
             ---> 1,2,3,4,7,8   2,7
                                               3,7

Figure 2. Observed Microbial Dechlorination Pathways for Highly Chlorinated Dioxins in Sediments and Sediment-Derived Microbial Systems.

3.2. BIOAVAILABILITY AND TRANSFORMATION KINETICS

The main underlying assumption with regard to bioavailability is that the microbially-mediated processes take place in the aqueous phase. Since hydrophobic contaminants such as PCDD/F have soil/water partitioning coefficients (K_d) on the order of 10^4-10^6 L.kg^{-1}, they will be mainly associated with the sediments, depending on the organic matter content. Bioavailability is then expected to be dependent on the mass transfer limited desorption rate of the PCDD/F from the sediments. Thus, if the rate of PCDD/F desorption is much slower than the rate of biodegradation, desorption will be the rate-limiting process for degradation, assuming that degradation takes place in the aqueous phase.

Whereas no information is available on desorption rates of PCDD/F from sediments, a biphasic kinetic desorption process has been observed with PCBs in both spiked and historically contaminated sediments. Using Saginaw Bay (MI) sediments, Di Toro and Horzempa [23] observed both a reversible and a resistant fraction of sediment-adsorbed hexachloro-biphenyl during sorption/desorption experiments. It was found that previously contaminated river and lake sediments required long desorption times, and that the lesser chlorinated PCBs desorbed significantly faster than congeners with a higher degree of chlorination [19, 5]. Adriaens et al. [3] developed the following relationship between the extractability of PCDD/PCDF from both high and low organic carbon sediments over a three year period, and the bulk partitioning coefficient Kd:

$$k_x \ (x \ 10^{-3} \ d^{-1}) = 0.3 \ \log K_d - 0.16 \qquad (r^2 = 0.91)$$

The issue of bioavailability of PCDD/F in complex soil-PCDD/F mixtures has been investigated with respect to the toxicity of sorbed 2,3,7,8-TCDD. It was found that biological uptake from these matrices was correlated with TCDD extractability, and thus soil composition [42]. Recently, Barkovskii and Adriaens [9] demonstrated that bioavailability of 2,3,7,8-TCDD may be linked to partitioning rate limitations in microbial membranes rather than to desorption rate limitations. Analysis of microbial cells derived from historically-contaminated sediments showed that the 2,3,7,8-substituted congeners had selectively partitioned into the microbial cell membranes, and that dechlorination could be stimulated. The rates of appearance of lesser chlorinated congeners in these cells was faster than that in freshly spiked microbial cells.

4. Reactivity of Dioxins and Pattern Recognition

As virtually nothing is known regarding the microbiology of reductive dechlorination in general, and that of dioxin dechlorination in particular, no conclusive biochemical rationale can be provided to explain the differences between treatments. Molecular calculations may then provide information to interpret the pathways observed, assuming no enzymatic or microbial preferences for specific substituent patterns is inferred. In the case of reductive dechlorination of aryl halides, good correlations have been obtained between predicted pathways based on molecular redox potentials or Gibbs Free Energy of formation and experimental observations [20,37,41]. Increasingly frequently, calculations of HOMO (Highest Occupied Molecular Orbital) - LUMO (Lowest Unoccupied Molecular Orbital) Gaps are used to describe molecular reactivities, whereby susceptibility to a redox reaction increases with decreasing potential differences (dE, eV) between HOMO and LUMO [31].

Based on *ab initio* calculations (RHF/6-31 G**) for octa to hexaCDD, a general trend emerged which indicated an increase in reactivity with increasing chlorine number, and increasing lateral chlorine substitution (Figure 3). This implies that, whereas the 2,3,7,8-chlorines confer stability on the PCDD molecule, *peri*-chlorines can be relatively easily removed resulting in an enrichment of 2,3,7,8-substituted residues. These calculations agree well with experimental observations where, in all non-pasteurized incubations, the 2,3,7,8-substituted hepta- and hexaCDD congeners are sequentially *peri* -dechlorinated. It should be noted that there may be a biochemical reason as well for this observation; based on the information obtained from historically-contaminated communities, the 2,3,7,8-substituted isomers will selectively partition into

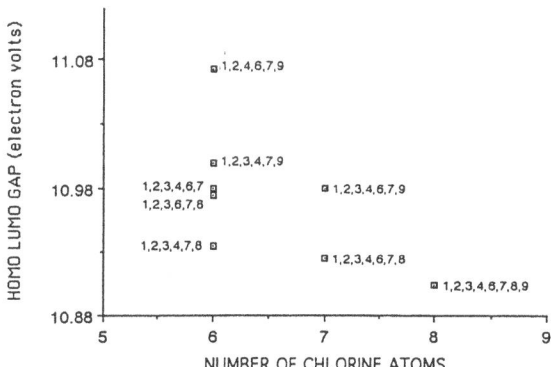

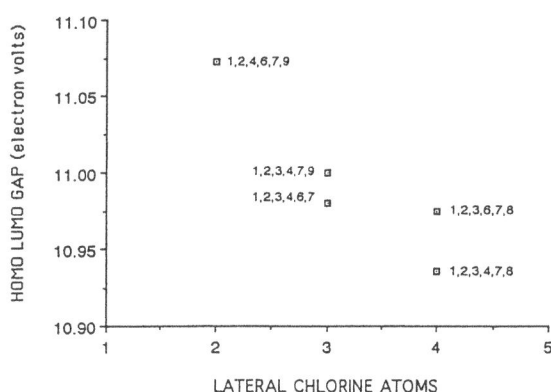

Figure 3. Results from Ab-Initio (RHF/6-31 G**) HOMO-LUMO gap calculations for selected PCDD congeners, observed to be formed during microbial reductive dechlorination

cell lipids, and potentially have the greatest potential for interaction with membrane-associated electron transfer components. Yet, based on these calculations alone, the relative predominance of the mixed *peri*-lateral dechlorination pathway can not be explained, and neither can the dependence of dechlorinating activity on the respiratory activity or humic acid interactions.

5. Conclusions and Future Recommendations

It is clear that a great many factors affect and may be able to effect PCDD dechlorination in reduced environments. Microbial activity appears to play either a primary (direct involvement) or secondary (indirect involvement by effecting changes in the sediment biogeochemistry) role in mediating these dechlorinations, and the ensuing patterns. The degree to which different microbial, molecular, or environmental parameters effect differences in the complex dioxin residues observed can not be evaluated using QSAR techniques alone.

The application of complex statistical methods for the recognition of chemical fingerprints is becoming an increasingly important tool to describe similarities and differences in sediments and biota containing complex mixtures of organic contaminants. Multivariate statistical analyses in particular are useful for addressing (i) data quality issues, (ii) identifying similarities in groups of samples, (iii) determining significant constituents among different samples, and (iv) providing a classification method for identifying the group or class to which a sample belongs. Yet, these techniques have found little or no application in biodegradation studies to identifying significant differences between treatments or reaction products resulting from these treatments.

Thus, our future efforts aimed at elucidating and interpreting complex chemical patterns in industrial waterways and heavily industrialized urban areas will include a combination of QSARs based on molecular descriptors, and information on microbial activity pertinent to the specific sediment biogeochemistry. Multivariate statistical analysis such as Principal Component Analysis (PCA) and Projection to Latent Structures (PLS) will need to be employed to evaluate the usefulness of the different molecular descriptors to explain the differences between experimental observations of reductive dechlorination reactions. Ultimately then, databases on source-sink analyses for dioxin residue distributions in anoxic sediments should be coupled to those containing information on microbial and abiotic dechlorination fingerprints (and the factors affecting them), such that dechlorination pattern-microbial activity relationships can be developed. These relationships will not only help resolve the interpretation of PCDD residues in sediments, but moreover become an important tool for long-term sediment risk assessment evaluations for 2,3,7,8-substituted congeners.

6. References

1. Adriaens, P., Barkovskii, A.L., and Albrecht, I.D. (1996). Fate of chlorinated aromatic compounds in soils and sediments, in D.C. Adriano, J.-M. Bollag, W.T. Frankenburger, and R. Sims (Eds), *Bioremediation of Contaminated Soils*, Soil Science Society of America/American Society of Agronomy Monograph, Soil Science Society of America Press, Madison WI. In Press.

2. Adriaens, P., Chang, P.R., and Barkovskii, A.L. (1996) Dechlorination of chlorinated PCDD/F by organic and inorganic electron transfer molecules in reduced environments. *Chemosphere* **32**, 433-441.

3. Adriaens, P., Fu, Q., and Grbic'-Galic', D. (1995) Bioavailability and transformation of highly chlorinated dibenzo-p-dioxins and dibenzofurans in anaerobic soils and sediments. *Environ. Sci. Technol.* 29, 2252-2261.

4. Adriaens, P., and Grbic'-Galic', D., (1994) Reductive dechlorination of PCDD/F by anaerobic cultures and sediments. *Chemosphere* **29**, 2253-2259.

5. Baker, J.E., and Eisenreich, S.J. (1990) Concentrations and fluxes of polycyclc aromatic hydrocarbons and polychlorinated biphenyls acros the air-water interface of Lake Superior. *Environ. Sci. Technol.* **24**, 342-352.

6. Barkovskii, A.L., and Adriaens, P. (1996) Extensive Microbial Reductive Dechlorination of Poly-chlorinated Dibenzo-p-Dioxins. *Appl. Environ. Microbiol.*, In Review.

7. Barkovskii, A.L., and Adriaens, P.(1996) Reductive dechlorination of polychlorinated dibenzo-p-dioxins: 1. Model humic constituents. *Environ. Sci. Technol.*, In Review.

8. Barkovskii, A.L, and Adriaens, P. (1996) Reductive dechlorination of polychlorinated dibenzo-p-dioxins: 2. Alteration of microbial activity by model humic constituents. *Environ. Sci. Technol.*, In Review.

9. Barkovskii, A.L., and Adriaens, P. (1995) Reductive dechlorination of tetrachorodibenzo-p-dioxin partitioned from historically-contaminated sediments to an authochthonous microbial community. *Organohalogen Compounds* **24**, 17-21.

10. Bedard, D.L., Bunnell, S.C., and Smullen L. A., (1996) Stimulation of microbial para- dechlorination of polychlorinated biphenyls that have persisted in Housatonic River sediment for decades. *Environ. Sci. Technol.* **30**, 687-694.

11. Bedard, D.L., and Quensen III, J.F. (1995) Microbial reductive dechlorination of polychlorinated biphenyls, in Young, L.Y.; Cerniglia, C.E., (eds) *Microbial Transformation and Degradation of Toxic Organic Chemicals*, Wiley-Liss, New York NY, pp. 127-217.

12. Beurskens, J.E.M, Toussaint, M., de Wolf, J. D, van der Steen, J. M. D, Slot, P.C., Commandeur, L.C.M., and Parsons, J. H. (1995) Dehalogenation of chlorinated dioxins by an anaerobic microbial consortium from sediment. *Environ. Toxicol. Chem.* **14**, 939-943.

13. Bonin, P., Ranaivoson, E.R., Raymond, N., Chalamet, A., and Bertrand, J.C. (1994) Evidence for denitri-fication in marine sediment highly contaminated by petroleum products. *Mar. Pol. Bull.* **28**, 89-95.

14. Brown, J.F., Jr., and Wagner, R.E. (1990) PCB Movement, dechlorination, and detoxication in the Auschnet estuary *Environ. Toxicol. Chem.* **9**, 1215-1220.

15. Carlton, R.G., and Klug, M.J. (1990) Spatial and temporal variations in microbial processes in aquatic sediments: Implications for the nutrient status of lakes, in R. Baudo, J. Giesy, and H. Muntau (eds), *Sediments: Chemistry and Toxicity of In-Place Pollutants*, Lewis Publishers, Inc., Ann Arbor, MI 1990, pp. 107-127.

16. Champine, J.E., Quensen III, J.F., and Tiedje, J. M. (1995) Differences in congener specificity of PCB-dechlorinating microbial communities obtained by indirect selection. *Abstr. Research Findings on Environmental Remediation and Toxicology*, Michigan State University pp.57-59.

17. Choudhary, G., Keith, L.H., and Rappe, C. (1983) *Chlorinated Dioxins and Dibenzofurans in the Total Environment;* Butterworth Publishers, Boston, MA.

18. Christian, R.R., Stanley, D.W., and Daniel, D.A. (1984) Microbiological changes occurring at the fresh-water-seawater interface of the Neuse River Estuary, North Carolina. in V.S. Kennedy (ed.) *The Estuary as a Filter*, Academic Press, Orlando FL, pp. 349-365.

19. Coates, J.T., and Elzerman, A. W. (1986) Desorption kinetics for selected PCBcongeners from river sediments. *J. Contam. Hydrol.* **1**, 191-210.

20. Cozza, C.L., and Woods, S.L. (1992) Reductive dehclorination pathways for substituted benzenes: a correlation with electronic properties. *Biodegradation* **2**, 265-278.

21. Czuczwa, J.M., and Hites, R.A. (1984) Environmental fate of combustion-generated polychlorinated dioxins and furans. *Environ. Sci. Technol.* **18**, 444-450.

22. De Angelis, M.A., and Scranton, M.I. (1993) Fate of methane in the Hudson River and Estuary. *Global. Biogeochem. Cycles* **7**, 509-523.

23. Di Toro, D.M., and Horzempa, L.M. (1982) Reversible and resistant components of PCB adsorption-desorption: Isotherms. *Environ. Sci. Technol.* **16**, 594-602.

64

24. Dutta, S.K., Hou, L.-H , Champine, J.E., and Tiedje, J.M. (1995) PCR-Based molecular analysis of meta and para dechlorinating microbial consortia. *Abstr. Research Findings on Enviornmental Remediation and Toxicology,* Michigan State University pp. 60-62.
25. Ehrlich, R., Wenning, R.J., Johnson, G.W., Su, S.H., and Paustenbach, D.J. (1994) A mixing model for polychlorinated dibenzo-p-dioxins and dibenzofurans in surface sediments from Newark Bay, New Jersey using polytopic vector analysis. *Arch. Environ. Contam. Toxicol.* **27**, 486-500.
26. Henrichs, S.M., and Reeburgh, W.S. (1987) Anaerobic mineralization of marine sediment organic matter: Rates and the role of anaerobic processes in the ocean carbon economy. *Geomicrob. J.* **5**, 191-237.
27. Kapone, D.G., and Kiene, R.P. (1988) Comparison of microbial dynamics in marine and freshwater sediments: Contrasts in anaerobic carbon metabolism.. *Limnol. Oceanogr.* **33**, 725-749.
28. Kostka, J.E., and Nealson, K.H. (1995) Dissolution and reduction of magnetite by bacteria. *Environ. Sci. Technol.* **29**, 2535-2540.
29. Lovley, D.R. (1991) Dissimilatory Fe(III) and Mn(IV) reduction. *Microbiol. Rev.* **55**, 259-287.
30. Näf, C., Broman, D.H. Pettersen, H., Rolff, C., and Zebühr, Y. (1992) Flux estimates and pattern recognition of particulate polycyclic aromatic hydrocarbons, polychlorinated dibenzo-p-dioxins, and dibenzofurans in the waters outside various emission sources on the Swedish Baltic Coast. *Environ. Sci. Technol.* **26**, 1444-1457.
31. Nevalainen, T., and Kohlemainen, E. (1994) New QSAR models for polyhalogenated aromatics. *Environ. Toxicol. Chem.* **13**, 1699-1706.
32. Nowicki, B.L. (1994) The effect of temperature, oxygen, salinity, and nutrient enrichment on estuarine denitrification rates measured with a modified nitrogen gas flux technique. *Est. Coast. Shelf Sci.* **38**, 137-156.
33. Ofjord, G.D., Puhakka, J.A. and Ferguson, J.F. (1994) Reductive dechlorination of Aroclor 1254 by marine sediment cultures. *Environ. Sci. Technol.* **28**, 2286-2295.
34. Olie, K, Vermeulen, P.L., and Hutzinger, O. (1977) Chlorodibenzo-*p* -dioxins and chlorodibenzofurans are trace compoounds of fly ash and flue gas of some municipal incinerators in the Netherlands. *Chemosphere* **8**, 455-459.
35. Ollivier, B.; Caumette, P., Garcia, J.-L., and Mah, R.A. (1994) Anaerobic bacteria from hypersaline environments. *Microbiol. Rev.* **5**, 27-38.
36. Painchaud, J.; Therriault, J.-C.; Legendre, L. (1995) Assessment of salinity-related mortality of freshwater bacteria in the Saint Lawrence Estuary. *Appl. Environ. Microbiol.* **61**, 205-208.
37. Peijnenburg, W.J.G.M., t'Hart, M.J., den Hollander, H.A., van de Meent, D., Verboom, H.H., Wolfe, N.L. (1992) QSARs for predicting reductive transformation rate constants of halogenated aromatic hydrocarbons in anoxic sediment systems. *Environ. Toxicol. Chem.* **11**, 310-314.
38. Prieur, D.,Troussellier, M., Romana, A., Chamroux, S., Mevel, G., and Baleux, B. (1987) Evolution of bacterial communities in the Gironde Estuary (France) according to a salinity gradient. *Est. Coast. Shelf Sci.* **24**, 95-108.
39. Rappe, C., Choudhary, G., and Keith, L.H. (1986) *Chlorinated Dioxins and Dibenzofurans in Perspective.* Lewis Publishers, Inc. Chelsea, Michigan.
40. Skyring, G.W. (1987) Sulfate reduction in coastal ecosystems. *Geomicrobiol. J.* **5**, 295-374.
41. Tratnyek, P.G., Hoigne, J., Zeyer, J., and Schwarzenbach, R.P. (1991) QSAR analyses of oxidation and reduction rates of environmental organic pollutants in model systems. *Sci. Tot. Environ.* **109/110**, 327-341.
42. Umbrett, T.H., Hesse, E.J. and Gallo, M.A. (1986) Bioavailability of dioxin in soil from a 2,4,5-T manufacturing site. *Nature,* 497-499.
43. Wenning, R.J., Paustenbach, D.J., Harris, M.A., and Bedbury, H. (1993) Principal component analysis of potential sources of polychlorinated dibenzo-p-dioxin and dibenzofuran residues in surficial sediments from Newark Bay, New Jersey. *Arch. Environ. Contam. Toxicol.* **24**, 271-289.
44. Wenning, R.J., Harris, M.A., Finley, B., Paustenbach, D.J., and H. Bedbury. (1993) Application of pattern recognition techniques to evaluate polychlorinated bibenzo-p-dioxin and dibenzofuran distributions in surficial sediments from the lower Passaic River and Newark Bay. *Ecotoxicol. Environ. Safety* **25**, 103-125.
45. Wenning, R.J., Harris, M.A., Ungs, M.J., Paustenbach, D.J., and Bedbury, H. (1992) Chemometric comparisons of polychlorinated dibenzo-p-dioxin and dibenzofuran residues in surficial sediments from Newark Bay, New Jersey and other industrialized waterways. *Arch. Environ. Contam. Toxicol.* **22**, 397-413.

A BIODEGRADABILITY EVALUATION AND SIMULATION SYSTEM (BESS) BASED ON KNOWLEDGE OF BIODEGRADATION PATHWAYS

BILL PUNCH and ARNOLD PATTON
Intelligent Systems Lab, A714 Wells Hall, Michigan State University
East Lansing, MI 48824 USA

KATHY WIGHT
Center for Microbial Ecology, 540 Plant & Soil Science,
Michigan State University, East Lansing, MI 48824 USA

BOB LARSON
The Procter & Gamble Company, Environmental Safety Department,
Ivorydale Tech Center, 5299 Spring Grove Ave, Cincinnati, OH 45217
USA

PATRIK MASSCHELEYN
European Technical Center, The Procter & Gamble Company,
Ternselaan 100 B-1853 Strombeek-Bever, BELGIUM

and

LARRY FORNEY
Center for Microbial Ecology, 540 Plant & Soil Science,
Michigan State University, East Lansing, MI 48824 USA

1. Abstract

BESS is a software system that simulates the action of biodegradation pathways on compounds. It does so by encoding biodegradation pathways in a knowledge base and applying those pathways in sequence to the compound, breaking it down into metabolites. We describe BESS, its knowledge-base, some preliminary validation results and an approach to learning that will be used to improve the knowledge base.

2. Introduction

The ability to predict the biodegradability of chemical structures under particular conditions is an important problem facing environmental science. The ability to make such predictions enhances the development of new chemicals since biodegradability can then more easily become an integral part of the design process itself.

There are a number of approaches to making such predictions. Our work attempts to combine two categories of these approaches. One class of work has used examples of known chemicals and attempts to determine the relationship between

65

W. J. G. M. Peijnenburg and J. Damborský (eds.), Biodegradability Prediction, 65–73.

biodegradation and other physics-chemical properties of a compound through the use of quantitative structure activity relationships via QSAR [1, 2], expert systems [3], statistical models [4, 5] or some other "learning" approach. The other class of work attempts to document the basic foundations of biochemical pathways for biodegradation.

BESS, a Biodegradability Evaluation and Simulation System is a project that has been ongoing for the last four years as a collaborative effort between a number of groups. The primary collaboration is one between Michigan State University and Procter & Gamble to develop BESS. Furthermore, the BESS system is the result of collaboration within Michigan State between members of the Intelligent Systems Lab and The Center for Microbial Ecology. The main goals for the development of this system are:

- Examine the biodegradation literature and collect into a compact, summary form, a list of known biodegradation pathways. This constitutes a set of "first principles" of biodegradation, including both general degradation pathways and degradation pathways important to P&G. These first principles include both the biological and environmental conditions necessary for these pathways to operate.
- Create a software system, BESS, that applies this "first principles" knowledge base to biodegradation of naturally occurring and xenobiotic compounds and predict a plausible pathway by which a compound could be metabolized, based primarily on the chemical's structure, the assumed environmental conditions and the available pathways.
- Integrate the BESS system with developing databases of biodegradation information. BESS could then operate in a mode of predicting biodegradation products until it finds information available on those products in the database. This allows the simulation to be "grounded in reality" whenever possible.
- Examine how BESS could be made to learn new biodegradation information so as to increase the knowledge base of biodegradation first principles, based on available experimental results.

The following Sections describe: the knowledge-base, BESS itself, and some preliminary experiments on adding learning to BESS.

3. The Knowledge Base

The collection of biodegradation knowledge was one of the most important, and one of the most difficult, tasks. While the most general pathways (β-oxidation, ω-oxidation, hydrolysis, etc) were well documented, it was relatively common to ask the question "How does the biodegradation of compound X occur", and not be able to find an answer. Moreover, it was often the case that when an answer was available, it was either in conflict with other literature sources or the conditions under which the pathway was observed were so specific (or, even worse, the conditions were not re-ported at all) that it required enormous effort to even partially validate pathways for some structures.

Nonetheless, these efforts led to the compilation of "first principles" on the biodegradation of various structures such as: Alkanes, Alkenes, branched Alkanes/Alkenes, Amino Acids (Leucine, IsoLeucine, Valine), Quaternary Alkanes,

Monocylic Hydrocarbons, Glyoxylates, Alkyl Ethoxlyates, Sulfonates and a number of other categories of structures. We documented general information about the biodegradability pathways useful for various structures, a description of the pathway(s), the conditions required for each pathway, and any conflicting information (and our resolution of conflicts when possible) based on multiple reference sources (over 300 references are used).

This "rule-book" was the basis for developing a knowledge base for the BESS system as described below.

4. The Structure of BESS

4.1. THE RULES

$$R_n - CH = CH - R_m \longrightarrow R_n - CH - CHOH - R_m \qquad n,m > 0$$

```
ruleName   := 'PathwayE::Rule1'
antecedent := [:aQuerier |
               aQuerier findMatchesFor: 'Xx.[C=C]1.xX'
                     withLeftAndRightFSCC:{:n :m | n=O and m>o}
               ]

consequent := [:aModifier |
               aModifier
                         designateTarget: I thru 2;
                         cut: 1 from: 2;
                         link: 1 to: 2 with: 1;
                         introduce: 'O' boundTo: 2 bondType: single;
                         identifyProducts.
               ]
```

Figure 1. An example of a rule for subterminal oxidation.

The information gathered in the knowledge-base was converted into rules of the following form. Each rule contains an antecedent-consequent pair. The antecedent describes a pattern sought in the chemical examined. More specifically, the antecedent encodes a search through the structure of the chemical for certain features. Consider the rule, shown in Figure 1, for subterminal oxidation of a double bond.

This code, written in Smalltalk^{TM}, is actually a small program. This antecedent code searches through the structure for a pattern containing a C=C bond, followed and preceded by a least one atom, where the atoms to the right and left of this pattern are fully saturated carbon chains. If this pattern is found, then the consequent of the rule is activated. First, the atoms of the structure are numbered, left to right, starting at the discovered pattern. These numerical designations are used by the consequent to perform a kind of "cut and paste" operation on the appropriate chemical bond to yield a metabolite. In the rule shown, the double bond is removed, replaced with a single bond, and an OH added to carbon 2 of the double bond pair.

4.2. RULE ORGANIZATION

The present BESS consists of approximately 100 rules. Because the antecedent pattern match operation can be quite time consuming, it is necessary to organize the rules such that only the rules pertinent to the present chemical are run. While this cannot always be the case, the closer one can get to this goal, the faster the system will run.

Presently, the rules are *organized* into groups, and these groups are organized into a *hierarchy*. Each group consists of three elements:

1. An entry or guard condition. This condition is a minimal requirement for *any* of the rules in this group to be applicable. BESS, while it is running, can eliminate whole groups of rules from examination if this guard condition cannot be met by the present chemical structure.

2. A list of rules. These rules are ordered within the group and, when activated, are searched in this order. The qualifications for grouping of the rules are essentially that they inter-operate, that is they typically work together to achieve some larger end. These constitute what is referred to as a biochemical module. For example, the β-oxidation group consists of about 6 rules which, when used in iteration, will remove as many terminal two-carbon units from a carbon chain as possible.

3. An exit condition, with a possible suggestion as to which group to activate next. Once a group is activated, BESS searches for rules to run *only* within the group rules. Thus an exit condition is required to get BESS to begin its search starting with the entire knowledge base (as opposed to some particular part of the knowledge base). Further, the exit condition can directly activate another group based on the exit conditions.

The hierarchical structure of the rule base is not significant at the moment, since all groups are essentially children of the same base node. However, we believe significant improvements in performance, both in terms of speed and accuracy, can be achieved through a better hierarchical organization (see Section 6 for more details). This is an area of active research.

4.3. BESS OPERATION

Figure 2 gives a broad overview of BESS's structure. BESS consists of the knowledge base (as a hierarchy of rule groups), a rule interpreter, an interface to external databases and a report generator. BESS begins with an input compound and begins to apply rules to modify that compound. It selects which rule group to use by searching the hierarchy top-down, looking for a rule group whose guard condition can be met. Once found, BESS begins running rules within that group until the exit condition is met. Upon exiting, if a next group is indicated by the exit condition itself, BESS tests the guard condition of the indicated group and, if possible, begins cycling within that group. If no group is indicated by the exit condition (or if the group indicated cannot be entered), then BESS begins searching again from the top of the knowledge base hierarchy.

At the start of the run, the initial compound structure is placed in the *structure list,* a list of structures that BESS must attempt to degrade. Before BESS begins simulation or rule-application, it queries any available databases to obtain any degradation information on the compound. Since BESS deals with compounds strictly via structure, it queries these databases on this basis. If the database has information on biodegradability of that structure, BESS halts and provides that database information in the final report. If not, BESS begins rule application. After applying a rule, BESS modifies the compound structure and produces metabolites. Both the modified structure and the products go back onto the structure list. Each structure in the list is treated as was the initial compound: the database is queried to determine if there exists any biodegradation information on that structure. If so, that compound is removed from the list and the database information used in the final report. If not, BESS will eventually process it: applying rules, modifying the structure and producing products. This process continues until the list of structures is emptied or no rules apply.

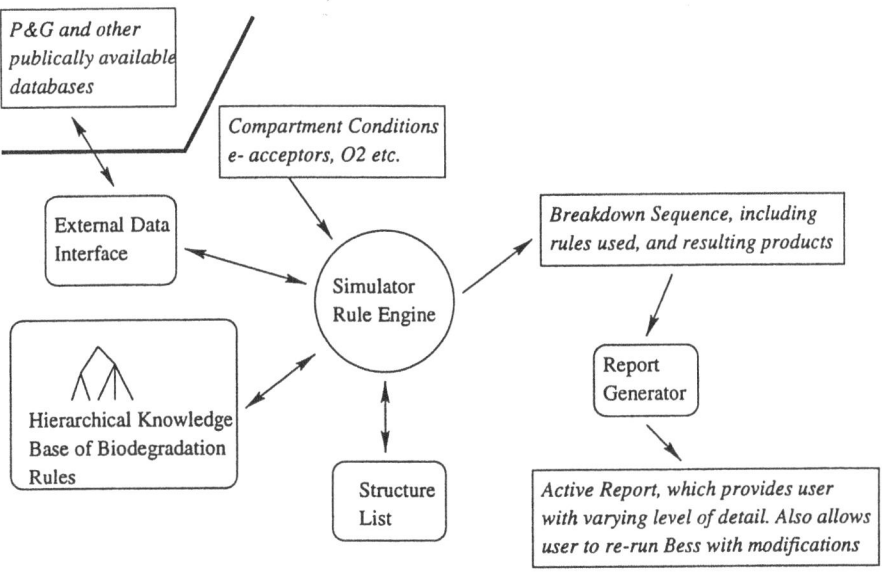

Figure 2. A broad overview of BESS's structure.

The result is a sequence of rule applications with associated structure modifications and resulting products. These are collected by the report generator and presented to the user. This report is an *active* report, that is the level of detail it presents can be determined by the user. The level of detail ranges from a simple "was or was not mineralized", to a complete listing of the result of each rule application and the conditions that activated each rule. In addition, this report allows the user to further explore other possibilities of degradation. For example, the user can re-run the simulation, starting from anywhere in BESS's rule application sequence, and modify the conditions, the rules or any other aspect of the simulation to play out various "what would happen if" scenarios.

Finally (not shown in the figure), a rule editor is used to update and add to the knowledge base. Each rule has fields for conditions of application, references, associated rule group and comments, as well as the required antecedent and consequent and the rule-editor aids in filling in these slots.

4.4. IMPLEMENTATION

BESS is implemented in VisualWorks™ Smalltalk. Smalltalk was the language chosen for two reasons. The first is that Smalltalk is a good language for initial development of software as its object-oriented nature and extensive object, graphics and interface libraries make prototyping much easier. Secondly, the VisualWorks environment is very portable between computing platforms. As a result, BESS can make heavy use of graphical user interfaces and still be completely portable across a number of systems (PC's, Mac's, Sun's, HP's).

5. Results

As BESS was developed we tested it on compounds with known degradation properties (both positive and negative), that is the compounds that were the basis for the rules. The next step in testing is to do a "blind-test" of BESS against an arbitrary test set of compounds. The difficulty in finding this test set is in picking compounds that the rules do indeed address (since we know that the rules are limited) without over-biasing the selection. Thus, we *know* that there are compounds that BESS cannot address since we have no rules in the knowledge-base that address such compounds. Moreover, we know that there will almost *always* be such compounds until a "critical-mass" of underlying rules are available in the knowledge-base. Nonetheless, we are in the process of testing BESS against a set of chemicals from the MITI book "Biodegradation and Bioaccumulation data of existing chemicals on the CSCL". It lists 431 compounds, 215 of which are readily degradable and 216 of which are not readily degradable.

Preliminary testing has shown that BESS can accurately predict the biodegradation of compounds when relevant biochemical pathway knowledge is available to the system. However, the BESS *knowledge base* is admittedly incomplete, and as a result of the validation testing we have identified three kinds of knowledge that BESS lacks:

-   *Base Cases:* BESS maintains a set of "base cases", structures which are considered end-points of degradation. If a degradation run creates one of the known base cases, that chemical is considered degradable because the base case is degradable. These base cases include compounds like acetic acid or succinate as well as other chemicals that enter into the central metabolic pathways of micro-organisims such as the tricarbo-oxylic acid cycle that are well known but not represented in BESS directly.

- *General Cases:* This is basically a deficiency of knowledge not yet covered in the rule base. It requires some analysis of the class of chemicals being worked on and the addition of rules to the knowledge base that are essential to simulate the degradation of a class of compounds. For example, in the early validation of BESS, no rules that apply to halogenated compounds were in the knowledge base, hence it was unable to accurately simulate the biodegradation of this class of compounds.
- *Special Cases:* When working on a compound never seen before, BESS often encounters a situation where it *could* give a full break-down sequence were it not for one *special* rule important for this (or its associated class) of compounds. In part this kind of problem can be remedied with the machine learning approach described in Section 6.

6. Learning, the Next Step

It is clear that the knowledge base is incomplete, and that knowledge acquisition will be required to cover an ever-expanding set of compound structures. What we have considered are ways to add learning to the BESS system. The question is, what kind of learning can we reasonably provide? We feel there are two areas where automated learning processes can be useful:

1. Learning a more "optimal" configuration of the rule hierarchy.
2. Learning new rules based on case studies.

The first approach is interesting both from a practical and scientific point of view. The practical aspect is that the rule organization, both the hierarchical structure of the rule groups and the constitution of rules within the groups, contribute to the efficiency of BESS. However, the resulting knowledge base structure is also interesting scientifically as it indicates something about the association between the pathways. If a certain group of rules is more efficient, or the hierarchical structure of the groups more efficient, it says something about how those rules work together, and which are more likely to be required.

The second approach is more difficult to define. It is relatively easy to write a system to learn rules based on the biodegradability of certain compounds. For example, (if Compound is X, then it is biodegradable) is not a very interesting rule. Nonetheless, *general* rules of biodegradability are important and difficult to learn. However, if an established knowledge base of biodegradation rules is available, it can serve as a resource for "guiding" such learning. Therefore, if the knowledge base does not properly predict the biodegradation of a known compound, we can *examine* the rules to determine where things went wrong. We can then find the *minimal* change required to make the prediction correct and then examine that change for any plausible suggestions based on the known biochemistry of biodegradation.

Consider a simple example. Studies show that compound X is biodegradable under conditions Y, but BESS predicts X is persistent. We could easily add a rule to handle X, but it would be much more useful to find where in the existing chain of rule application the problem occurred. Once found, we can then add a new rule, under the following conditions:

- The rule is *minimal*, in the sense that it is just enough to complete the rule application chain, but changes as little as possible otherwise.
- The rule "fits" in with existing knowledge. That is, it cannot contradict established rules nor violate established conditions unless those rules or conditions can be somehow proven to be incorrect.

This is obviously an iterative process. Proposing a new rule may affect other known "correct" solutions, so the new rule must be tested on these solutions to insure that incorrect knowledge (of microbial physiology and enzyme reaction mechanisms) was not added to fix the particular X problem. However, if such knowledge can be found, then it can suggest with some force that some new pathways or conditions necessary for biodegradation may exist.

6.1. GENETIC ALGORITHMS FOR LEARNING

Our approach to these learning problems is the use of a genetic algorithm to optimize the performance of the rule base. The structure of this approach is shown in Figure 3.

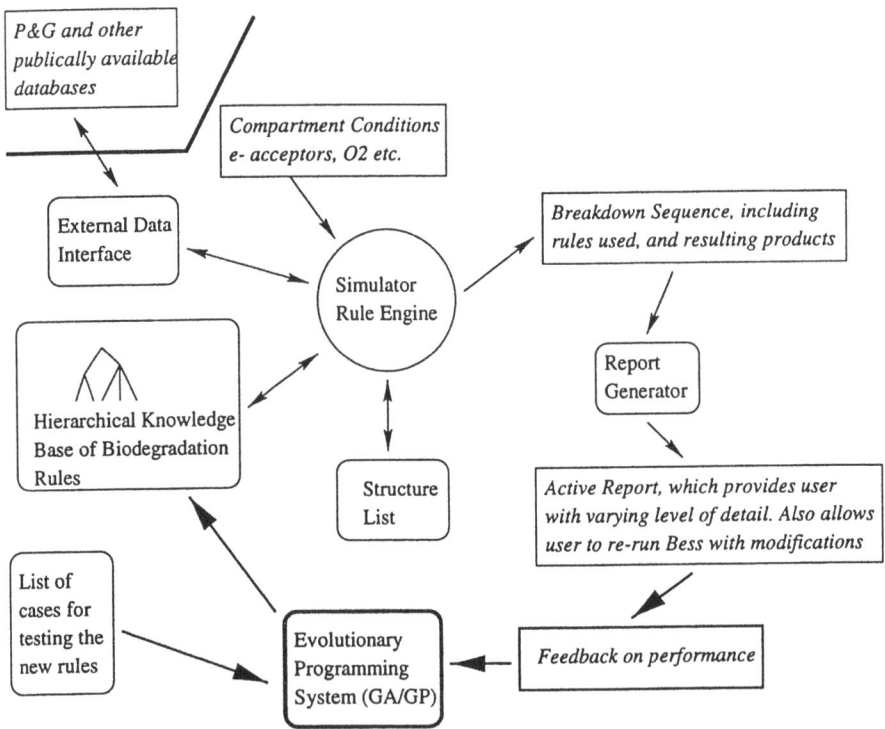

Figure 3. BESS as Modified for Learning.

A genetic algorithm (GA) is an optimization technique that operates on a *population* of individual solutions. Each individual solution, also called a *string* in the population, represents a proposed solution to the problem being solved. The theories of natural selection are applied to this population, and subsequent generations of the population are obtained by applying selection, reproduction, and mutation operators (among possible others) to the population. With these operators, the population of solutions is gently pushed towards a good- and hopefully the optimum-solution to the problem.

The GA designer provides a fitness function to evaluate the fitness of each individual solution; this fitness function is used to propagate "good" individuals into the next generation. Some set of these fit individuals are chosen for a *crossover* operation, which recombines the strings of the parents into new children, trying to build up healthier strings in the process. The mutation operator randomly alters some element of an individual in order to further enhance the population.

For our approach, each string or solution would constitute a complete knowledge base, that was derived through modifications (additions, deletions or corrections) to the original knowledge base. The fitness function would be the performance of that new knowledge base on known solutions and on new problems for which we are trying to find new rules. The result is an iterative process that is rather slow, but very capable in terms of finding new rules.

7. Future Directions and Requirements

The main objectives for the future development of BESS are the implementation of the learning system as described in Section 6 and to more fully develop the knowledge base available through the use of published studies on the biodegradation of various chemicals.

8. References

1. Hansch, C., Leo, A., and Hoekman, D. (1987) *Exploring QSAR Volume 1: Fundamentals and Applications in Chemistry and Biology*, ACS books, Washington, DC,
2. Sanz, F. (1995) *QSAR and Molecular Modeling 94*, Prous, Barcelona.
3. Pangrekar, J., Klopman, G., and Rosenkranz, H. (1994) Expert-System Comparison of Structural Determinants of Chemical Toxicity to Environmental Bacteria, *Environmental Toxicology and Chemistry*, **13**, 979-1001.
4. Niemi, G., Veith, G., Regal, R., and Vaishnav, D. (1987) Structural Features Associated with Degradable and Persistent' Chemicals, *Environmental Toxicology and Chemistry*, **6**, 515-527.
5. Howard, P., Boethling, R., Stiteler, W., Meylan, W., Hueber, A., Beauman, J., and Larosche, M. (1992) Predictive Model for Aerobic Biodegradability Developed from a File of Evaluated Biodegradation Data, *Environmental Toxicology and Chemistry*, **11**, 593-603.

A MECHANISTIC APPROACH TO DERIVING QUANTITATIVE STRUCTURE BIODEGRADABILITY RELATIONSHIPS

A Case Study: Dehalogenation of Haloaliphatic Compounds

J. DAMBORSKÝ, K. MANOVÁ AND M. KUTÝ
Faculty of Science, Masaryk University
Kotlárská 2, 611 37 Brno, *CZECH REPUBLIC*

1. Abstract

The application of a mechanistic approach for the study of mechanisms of microbial degradation processes and the development of Quantitative Structure-Biodegradability models are outlined in this contribution. The dehalogenation of haloaliphatic compounds was used as a case study and an attempt was made: (i) to determine the rate-limiting sub-process, (ii) to quantitatively estimate interspecies variability in substrate specificity and (iii) to investigate structure-activity relationships leading towards development of QSBR models.

A comparison of dehalogenation rates obtained in testing systems at different organization levels, intact cells and isolated enzyme, revealed that penetration of the halogenated aliphatic compounds into the cells of dehalogenase competent microbes is not the rate limiting step in the hydrolytic dehalogenation. Consequently, the enzymatic reaction is to be considered slower than the penetration process. Multivariate analysis of the substrate profiles of the haloalkane dehalogenases of different microbial strains was performed to tackle the problem of the utility of single species testing for the prediction of environmental rate constants. At least two groups of haloalkane dehalogenases were formulated based on their ability to dehalogenate various haloaliphatic compounds.

A preliminary QSBR model for terminally substituted mono- and dihalogenated alkanes was developed. All three types of descriptors: hydrophobicity, steric, and electronic were necessary to obtain a good description of the process studied. Three outliers from the model, all short-chain chlorinated compounds, were detected and the reasons for discrepancy between predictions and observation are discussed. Although more research is needed to better understand structure-biodegradability relationships for hydrolytic dehalogenation, the mechanistic approach is shown to provide beneficial information for model interpretation.

2. Introduction

The possibility to predict the biodegradability of organic compounds directly from their chemical structure is challenging and a number of attempts have been made to mathematically describe the relationships between the structure of organic compounds

75

W. J. G. M. Peijnenburg and J. Damborský (eds.), Biodegradability Prediction, 75–92.

and their biodegradability - the susceptibility to attack and decomposition by the action of (micro)organisms. A number of so called QSBR (Quantitative Structure-Biodegradability Relationships) models have been published in the scientific literature and many of them have been recently critically discussed in several review papers [1-4]. A general conclusion arising from these reviews is that the present state of the art in biodegradability predictions is not satisfactory. One of the factors limiting the further development of the new structure-biodegradability models certainly is the high complexity of the microbial degradation process and our limited understanding of its mechanisms. To a certain extent our knowledge of the mechanisms of biodegradation exist (e.g., [5-9]), but this knowledge is not utilized in the models.

The application of simple molecular descriptors and/or molecular fragments treated by advanced statistical methods can lead towards construction of robust models with good predictive ability). The relationships between the structure of studied compounds and their biodegradability as described in these models usually are not well understood which may limit their practical utility as well as constrain their further improvement.

The purpose of the mechanistic approach [10] to deriving structure-biodegradability relationships is to study the various mechanisms taking place during the degradation process and to investigate the possibilities of incorporation of these mechanisms in the QSBR models. The mechanistic approach proposes: "comparison of the various biological data measured with different species, under different conditions and at the different organization levels in order to extract and quantify the mechanisms taking place during the microbial degradation" and is intended to view and treat the biodegradation process as a biological one performed by living organisms with their

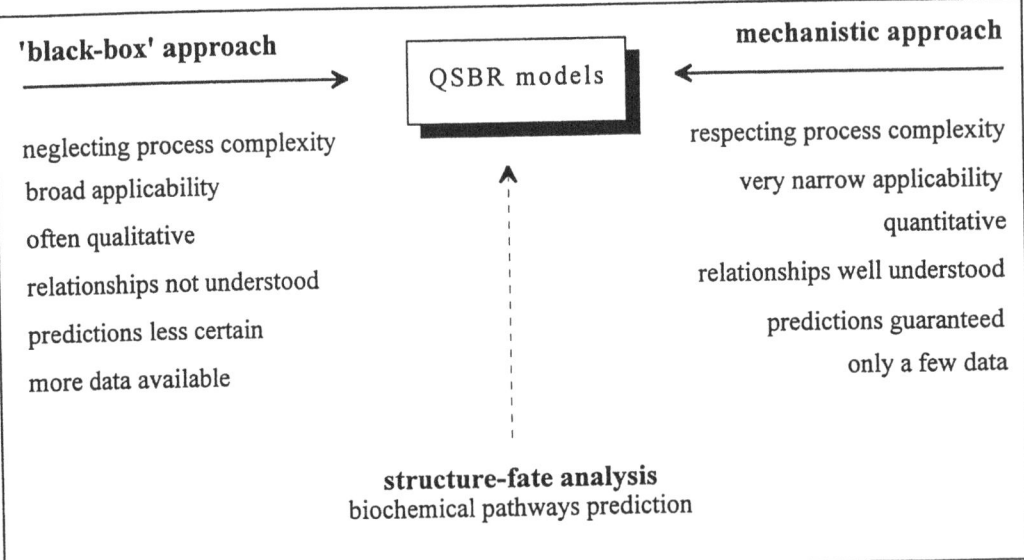

Figure 1. Combination of different approaches used for structure-biodegradability relationships can contribute towards development of the QSBR models with good predictive power and understood relationships.

specific physiology, genetics and enzymatic activity, surrounded by different environments. Of course, it is a long way to generalize the models obtained using mechanistic approach to such a extend, that these could be used for predictions at a environmental level, but it is a alternative to 'black-box' approach and the combination of both approaches may lead towards development of the new quality QSBR models (Figure 1). This contribution summarizes the results obtained to date from the application of the mechanistic approach to a specific case, namely the dehalogenation of haloaliphatics compounds.

3. Methods

3.1. BIOLOGICAL (DEHALOGENATION) DATA
Biological data describing the kinetics of dehalogenation reactions were obtained from assays using the biological system of different organization levels: (i) intact cells of the soil bacteria *Rhodococcus erythropolis* Y2, (ii) the crude extracts of the same strain *Rhodococcus erythropolis* Y2, (iii) the pure enzyme, haloalkane halidohydrolase, isolated and purified from *Rhodococcus erythropolis* Y2. Our experimental data have been complemented with dehalogenation data extracted from the literature [12-16,11,17-19].

3.1.1. Dehalogenation testing with (intact) resting cells
Resting cell assays have been performed with the biomass of 1-chlorobutane grown *Rhodococcus erythropolis* Y2 [11]. The cells have been grown at 30 °C on a reciprocating shaker (120 rpm) in the mineral medium containing (per litre of distilled water): 5.37 g $Na_2HPO_4.12H_2O$, 1.36 g KH_2PO_4, 0.2 g $MgSO_4.7H_2O$, 0.5 g $(NH_4)_2SO_4$, 20 mg yeast extract, 5 ml of trace elements solution according to Janssen [20], amended with 5mM of chlorobutane as the only carbon and energy source. Biomass has been harvested in the exponential phase by centrifugation 8000 x g and 4 °C for 30 min, washed twice in 100 mM glycine/NaOH buffer (pH 9.1) and resuspended in the same buffer to reach an optical density of 0.8 (600nm). Fresh biomass was prepared for each assay and spiked with tested compound (calculated concentration 5mM). A 'real' concentration was somewhat lower due to low solubility and high volatility of the halogenated aliphatic compounds. Chlorine ions released during the dehalogenation reaction were monitored in the samples withdrawn in periods 0, 1 and 6 hours using the spectrophotometric method of Iwasaki [21]. Linear formation of the product within a sampling period have been observed and the slope of a plot of time versus chlorine concentration has been used for quantitative expression of the dehalogenation rate. A correction has been made for abiotic dehalogenation and small changes in biomass during the assay period.

3.1.2. Dehalogenation testing with crude-cell free extracts
The detailed procedure used for the crude cell-free extracts preparation and kinetic dehalogenation measurements has been described elsewhere [22]. Briefly, the cells of *Rhodococcus erythropolis* Y2 grown on the minimal medium with 5mM 1-chlorobutane have been harvested in exponential phase by centrifugation, washed twice in phosphate buffer (pH 9.1), frozen overnight and then disrupted by three passages through a French pressure cell. Intact cells and debris were removed by centrifugation. The dehalogenation assay was performed in a glycine/NaOH buffer (pH 9.1), spiked with 5mM of

tested compound, using spectrophotometry for monitoring the release of halide ions in the samples withdrawn after 15, 30 and 45 min.

3.1.3. Dehalogenation testing with purified enzyme

The hydrolytic haloalkane dehalogenase of *Rhodococcus erythropolis* Y2 was purified from the crude extract using the protocol of Sallis et al. [11] with three chromatographic steps: (i) DEAE-Sephacel, (ii) hydroxyapatite and (iii) gel-filtration. The specific enzyme activity was increased 24-fold (TABLE 1) and the final preparation contained one major protein band on SDS-PAGE. The same method as described for crude cell-free extract (3.1.2.) was used in the kinetic measurements with pure haloalkane dehalogenase.

TABLE 1.   Purification of the haloalkane halidohydrolase of *Rhodococcus erythropolis* Y2

Purification step	Volume [ml]	Protein conc. [mg/ml]	Specific activity [μmol/min/mg]	Recovery [%]	Purification fold
Cell-free extract	105	1.291	101	100	-
DEAE	106	0.154	711	85	7
Hydroxyapatit	85	0.052	2063	67	20
Gel filtration	59	0.04	2452	44	24

3.2. CHEMICAL (DESCRIPTOR) DATA

3.2.1. Physico-chemical properties

Five physico-chemical descriptors have been used for the global characterization of the studied compounds: molecular weight (Mw), boiling point (bp), refractive index (n), density (D), octanol/water partition coefficient (logP), moments of inertia (IX, IY, IZ) and Sterimol (L, B1-4) constants. Partition coefficient values were calculated using the ClogP program [23], while the moments of inertia and Sterimol constants using the TSAR 2.2 program [24].

3.2.2. Quantum-chemical descriptors

Fifteen molecular and atomic quantum-chemical descriptors have been calculated for geometry optimized structures: heat of formation (Hf), total energy (TE), electronic energy (EE), ionization potential (IP), energy of the highest occupied molecular orbital (HOMO), the energy of the lowest unoccupied molecular orbital (LUMO), the charge on leaving halogen atom (QX), the charge on the carbon atom (QC), the dipole moment (Dip), the bond contribution of the highest occupied molecular orbital (BCHO), the bond contribution of the lowest unoccupied molecular orbital (BCLU), bond order (BO), nucleophilic delocalizability (Dn), absolute hardness (Hard), carbon-halogen distance (CX). Input structures were built in the Builder module of the InsightII [25] molecular modelling package and pre-optimized using its Discover molecular mechanic program with a cvff forcefield. Final optimization and molecular orbital calculations have been performed with the MOPAC 6.0 programme using a standard BFGS optimization algorithm [26].

3.3. STATISTICAL ANALYSIS

Multivariate statistical methods [27], namely Principal Component Analysis (PCA), Cluster Analysis (CA) and Projection to Latent Structures (PLS) have been used for data analysis. Simca-S 5.1b [28] implemented algorithms have been used for PCA and PLS analysis, while the TSAR 2.2 program was used for CA analysis [24]. Where appropriate, Cross-validation [29] has been applied in order to obtain good predictive power of the models developed. In most cases the biological data have been logarithmically transformed to improve their normal distribution.

3.4. MOLECULAR MODELLING

3.4.1. Input structures

The active site model used in the semiempirical calculations of the complete reaction pathway for dehalogenation of 1,2-dichloroethane by haloalkane dehalogenase of GJ10 consisted of 13 amino acid residues: Glu56, Asp124, Trp125, Phe128, Phe164, Phe172, Trp175, Phe222, Pro223, Val226, Asp260, Leu262, His289 and the substrate molecule 1,2-dichloroethane (DCE). Input co-ordinates of the enzyme-substrate complex of GJ10 haloalkane halidohydrolase soaked with 1,2-dichloroethane have been obtained from the Brookhaven Protein Database (entry 2DHC).

Three amino acid residues of the enzyme active site plus substrate molecule (1,2-dichloroethane) have been considered in simple active site models used for comparison of the first (SN2) reaction step as catalysed by haloalkane dehalogenases of *Xanthobacter autotrophicus* GJ10 and *Sphingomonas paucimobilis* UT26. The model of GJ10 enzyme active site consisted of Asp124, Trp125, Trp175 and DCE whilst the model of UT26 consisted of Asp124, Trp125, Phe175 and DCE (both numbered according to GJ10 sequence). Input co-ordinates of the enzyme-substrate complex of GJ10 have been obtained from the Brookhaven Protein Database (entry 2DHC). The active site of UT26 was modelled by substitution of Trp175->Phe followed by minimization using the cvff forcefield and steepest descent algorithm of Discover program [25].

3.4.2. Quantum chemical calculation of a reaction pathway

The subroutine DRIVER of the semi-empirical quantum chemical program MOPAC 6.0 [26] has been used for the mapping of the possible reaction pathway of the dehalogenation reaction catalysed by the haloalkane dehalogenase and for the comparison of the first reaction step performed by GJ10 and UT26 enzyme. The AM1 Hamiltonian and standard BFGS optimization algorithm have been used through the all calculations. The activation energy (Ea) and enthalpy difference (ΔH) values have been deduced from calculated energies of the educt, the product and the transition state.

4. Results and Discussion

4.1. SEARCH FOR THE RATE-LIMITING STEP

In structure-activity relationships one looks for the molecular descriptors which quantitatively describe those chemical properties which relate to activity. It is essential to choose such descriptors which are indeed relevant for the process under study. It has already been stated in the introduction that the biodegradation process is very complex and consists of a number of sub-processes, which include uptake (penetration) of the compound into the cells, enzymatic conversion, release of the products, etc. Each of these sub-processes can theoretically be rate limiting, depending on the chemical character and size of the compound, affinity of the enzymes towards the compound, the kinetics of the chemical reactions, etc. The determination of the rate-limiting step of the degradation process of interest is highly desirable since the descriptors used for structure-activity modelling usually have to relate to this particular step.

I. Dehalogenation rates of α,ω-chlorinated alkanes obtained in parallel with intact cells and purified dehalogenase enzyme have been compared (TABLE 2) in order to investigate whether or not penetration of haloalkanes into the degrading organism is the slowest (rate-limiting) step of the dehalogenation reaction. Very similar dehalogenation profiles (correlation coefficient R = 0.97, slope 1.061) have been observed for the rates measured in system with and without a cell envelope. This result indicate that penetration of even longer chain compounds, like 1,9-dichlorononane or 1,10-dichlorodecane, does not limit their rate of dehalogenation and that penetration is faster than enzymatic reaction. The single enzyme is responsible for the first dehalogenation step of studied compounds and the rate constant of this biochemical reaction will determine the overall kinetics of the dehalogenation by the strain *Rhodococcus erythropolis* Y2 under the conditions described in the Methods section.

TABLE 2.   Dehalogenation profile obtained on α,ω-chlorinated alkanes with intact cells and pure dehalogenase enzyme of *Rhodococcus erythropolis* Y2

compound	relative dehalogenation[1] [%]	
	intact (resting) cells[2] 2	pure enzyme[3]
1,2-dichloroethane	0	0
1,3-dichloropropane	112	139
1,4-dichlorobutane	125	130
1,5-dichloropentane	96	95
1,6-dichlorohexane	106	103
1,8-dichlorooctane	101	98
1,9-dichlorononane	99	101
1,10-dichlorodecane.	77	83

[1] relative values are expressed as a percentage of dehalogenation rate constant obtained under the same conditions with 1-chlorobutane
[2] values represent average of six replicates
[3] values represent average of nine replicates

II. Examination of the data of Armfield et al. [16] measured on a-chloroalkanes with (i) 1-chlorobutane and (ii) hexadecane grown resting cells of *Rhodococcus erythropolis* Y2 provided further evidence that penetration does not determine the rate of dehalogenation. A dramatic change in dehalogenating activity of the *Rhodococcus* cells grown on two different substrates was observed (Figure 2) and suggested induction of two different enzymes - hydrolase and oxygenase type dehalogenase, respectively [16].

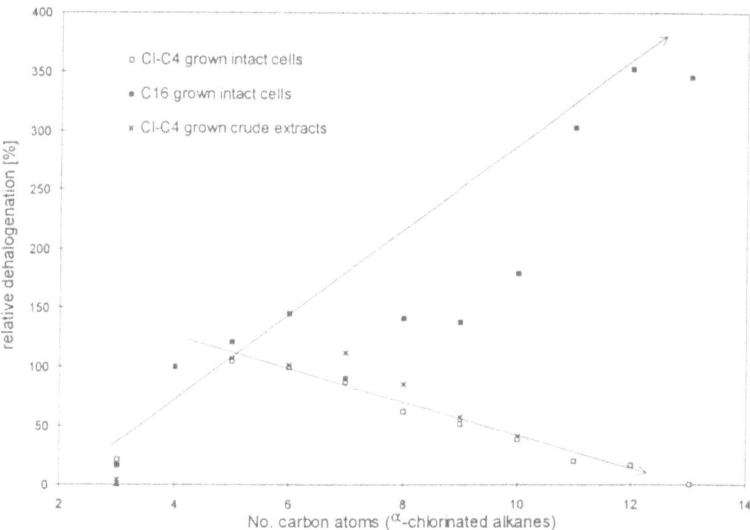

Figure 2. Dehalogenation profiles obtained with 1-chlorobutane (Cl-C4) grown intact cells and crude extracts, and hexadecane (C16) grown intact cells [16].

Chlorobutane grown cells and extracts with presumably induced haloalkane dehalogenase displayed decreasing dehalogenation rates with increasing number of carbon atoms in the substrate, while the opposite trend, improved dehalogenation activity with increasing chain length has been detected when hexadecane grown cells with induced oxygenase enzyme were used. If the penetration process is rate-limiting, the same or a very similar profile would be observed no matter which type of enzyme catalyses the reaction.

Both examples discussed above (I and II) demonstrate that the penetration of compounds through the membranes of dehalogenating organisms is not rate-limiting for the process under study. At the same time, the biochemical reaction seems to have fundamental influence on the dehalogenation profile. Further investigation of biochemical reaction(s) is needed to identify the rate-limiting step of hydrolytic dehalogenation. Biochemical catalysis performed by a particular enzyme can be divided into three or more additional sub-processes: (i) diffusion and binding of the substrate at the enzyme active-site, (ii) chemical reaction and (iii) release of the product. It can be very difficult to quantitatively describe these sub-processes using experimental techniques and the application of molecular modelling offers a possible alternative for further study.

A molecular modelling approach is particularly useful in cases when 3D structure of the target macromolecule is known or can be at least predicted using techniques, like receptor mapping [30-33] or homology modelling [34-37]. More extensive application of molecular modelling in the study of the mechanisms of degradation reactions at the 'active-site level' is limited by the number of known 3D structures of the enzymes involved in biodegradation pathways: cytochrome P450cam [38], protocatechuate 3,4-dioxygenase [39], haloalkane dehalogenase [40], methane monooxygenase [41], 2,3-dihydroxybiphenyl dioxygenase [42,43], L-2-haloacid dehalogenase [44], but it can be expected that even more structures will be solved using X-ray crystallography or NMR spectroscopy in near future.

Crystallographic studies of Verschueren et al. [45-48] leading towards determination of 3D structure of haloalkane halidohydrolase of *Xanthobacter autotrophicus* GJ10 and a proposal of the reaction mechanisms of dehalogenation reactions provided excellent input information for a detailed (quantitative) study of biochemical catalysis of dehalogenation and the search for its rate-limiting step. The reaction of haloalkane dehalogenase of *Xanthobacter autotrophicus* GJ10 proceeds via a nucleophilic attack of the enzyme aspartate on the carbon atom bearing a halogen leading to the formation of an alkyl-enzyme ester, which is subsequently hydrolysed by the water molecule present in the active site [45]. The energy profile of the SN2 reaction step of the wild type enzyme and its mutants were studied using quantum chemical tools [49].

Figure 3 shows the possible reaction pathway of the first two steps of the hydrolytic dehalogenation reaction catalysed by the haloalkane dehalogenase of *Xanthobacter autotrophicus* GJ10 as obtained from quantum-chemical AM1 calculations on a simplified model of the enzyme active-site (Glu56, Asp124, Trp125, Phe128, Phe164, Phe172, Trp175, Phe222, Pro223, Val226, Asp260, Leu262, His289 + 1,2-dichloroethane). An apparent limitation of the methodology used is that only those

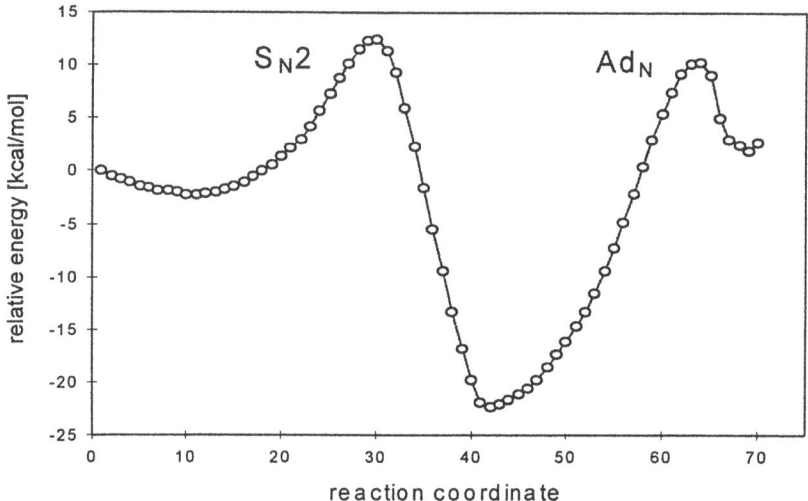

Figure 3. The proposed reaction pathway for the first two steps of the dehalogenation reaction catalysed by haloalkane dehalogenase of *Xanthobacter autotrophicus.*

enzyme residues which are in direct contact with the substrate during the reaction are included in the calculation, while the reminder of the protein is neglected as well as the *in-vacuo* nature of the calculation. Nevertheless, the inclusion of the rest of the protein and the solvent into the calculation, e.g. by using a hybrid QM/MM method [50,51] may change the absolute values of energies of reaction pathway but the relative differences in activation energies of the reaction steps are likely to be similar to our present result which suggests, that it is the second or third (not shown) reaction step (product release respectively) which is rate limiting. Additional calculations are needed to confirm this conclusion.

A molecular modelling study may be used in the search for the rate-limiting step, in addition new descriptors for QSBR analysis can be obtained when calculations are being performed with the model of one enzyme and many different compounds. Such descriptors, like activation energy or binding energy, may certainly be relevant to the biodegradation process under study [52]. In addition, the geometries of the transition states of the compounds obtained from the molecular modelling study can be used for calculation of quantum-chemical descriptors [53] which may better relate to the (bio)degradation reaction than those calculated for ground state geometries.

4.2. INTER-SPECIES DIVERSITY IN SUBSTRATE SPECIFICITY

A better knowledge of the variability in substrate specificity of degrading organisms and their enzymes may help to address the problem of the validity and applicability of QSBR models based on kinetic data from single species tests. The substrate specificity and absolute activity of the "key enzyme" of a particular degradation pathway directly determines the ability of a degrading species to attack and decompose xenobiotic compounds. Of course, there are many other factors of utmost importance, like efficient expression of the genes coding the degrading enzymes, the bioavailability of the compound of interest and its permeability via the biomembranes, the resistance of the degrading organisms towards the compound, etc.

The dehalogenation data [12-16,11,17-19] obtained from the tests with same group of halogenated aliphatic compounds and different haloalkane dehalogenases (crude extracts or pure enzymes) have been quantitatively compared using cluster analysis and principal component analysis [54,19]. All enzymes considered in the analysis, namely the dehalogenases of *Xanthobacter autotrophicus* GJ10 [12], *Pseudomonas* sp. E4M [17], *Acinetobacter* sp. GJ70 [15], *Rhodococcus* sp. HA1 [13], *Rhodococcus erythropolis* Y2 [11], *Rhodococcus* sp. m15-3 [14], *Corynebacterium* m2C-32 [55], *Rhodococcus* sp. CP9 [18] and *Sphingomonas paucimobilis* UT26 [19], are known to dehalogenate by a hydrolytic mechanism. The results indicate that at least two, but probably more groups of haloalkane dehalogenases can be distinguished Figure 4. The differences in dehalogenation rates among these groups can be very significant requiring that separate QSAR models need to be considered for each group. It may be concluded that the same reaction mechanism of biotransformation or biodegradation does not guarantee the same specificity profile as has been previously pointed out by Eriksson et al. [56].

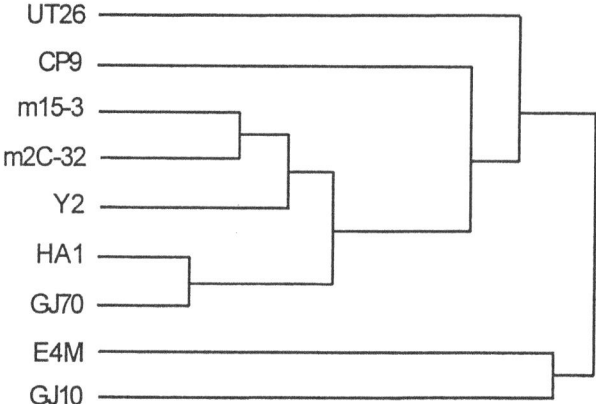

Figure 4. Dendrogram from cluster analysis of dehalogenation profiles [12-16,11,17-19] of the haloalkane dehalogenases isolated from different strains.

Structure-function relationships analysis for dehalogenase enzymes needs to be applied in order to understand the reasons for differences in specificity profiles that have been found, for example, between the GJ10 and UT26 haloalkane dehalogenases. Comparison of the X-ray structure of GJ10 dehalogenase [57] with the homology model of UT26 dehalogenase [36] revealed some presumably important differences in the active site composition of these two enzymes. An attempt has been made to investigate the influence of one of these differences, substitution of Trp175 by Phe (according to GJ10 sequence numbering), on the catalytic activity with 1,2-dichloroethane using quantum chemical calculations of the reaction pathway [49]. The significant change in kinetic and thermodynamic parameters of the first reaction step has been calculated for very simple models of the enzymes active sites (TABLE 3). This result provides one

TABLE 3. Quantum mechanic study of the influence of Trp175 -> Phe substitution on kinetic and thermodynamic characteristic of SN2 dehalogenation reaction step

active site model	residue 175[1] [%]	dehalogenation[2] kcal/mol	Ea[3] kcal/mol	ΔH[3]
GJ10	Trp	360	21	-12
UT26	Phe	0	28	-4

[1] residue number according to the GJ10 sequence
[2] percentage of dehalogenation rate constant obtained with 1-chlorobutane [12,19]
[3] energy barriers as obtained from MOPAC AM1 calculation [49]

possible explanation for the inactivity of *Sphingomonas paucimobilis* UT26 dehalogenase with 1,2-dichloroethane, which is opposite to the activity observed with the *Xanthobacter autotrophicus* GJ10 enzyme.

4.3. PRELIMINARY QSBR MODELLING OF HYDROLYTIC DEHALOGENATION
An attempt to develop structure-biodegradability model was made in parallel with a detailed study of the dehalogenation process. Neither the rate-limiting step nor all mechanisms important for the process have been recognized to date, therefore, a large

number of descriptors have been generated for the studied haloaliphatic compounds and multivariate statistics applied for selection of those which are relevant for modelled processes.

To date the first run of analysis for data retrieved from the literature (not shown) revealed that it will be rather challenging if not impossible to find a general QSBR model for all the data. The reason for this is that the chemicals had been selected by authors for testing without systematic variation of the molecular structure resulting in a data matrix with a mixture of several chemical species: terminally halogenated alkanes, β-substituted haloalkanes, chlorinated alcohols, chlorinated carboxylic acids and chloroalkenes, in which the latter groups were represented by only a few compounds. PLS analysis of the data matrix of such composition then resulted in a single significant component, explaining one systematic structural variation in data - usually the length of the carbon chain - and leaving many compounds as outliers. Analysis of small groups of

TABLE 4.   Dehalogenation rates and descriptors used in Model 1

chemical name	Y2[1]	MW	IX	logP	Hf	TE	EE	HOMO	Dip	BCLU
1-chloropropane	21	78.54	3.28	2.04	-33.01	-855.16	-2527.2	-11.135	1.73	-1.978
1-chlorobutane	100	92.57	4.97	2.64	-39.82	-1010.99	-3438.5	-11.133	1.74	-1.979
1-chloropentane	105	106.6	5.67	3.05	-46.67	-1166.82	-4424.7	-11.133	1.76	-1.968
1-chlorohexane	99	120.62	7.18	3.58	-53.51	-1322.66	-5475.5	-11.133	1.77	-1.981
1-chloroheptane	87	134.65	7.94	4.15	-60.36	-1478.49	-6581.83	-11.133	1.78	-1.981
1-chlorooctane	62	148.68	9.37	4.64	-67.23	-1634.33	-7737.5	-11.133	1.78	-1.981
1-chlorononane	51	162.71	10.16	5.17	-74.05	-1790.16	-8937.1	-11.133	1.79	-1.981
1-chlorodecane	38	176.73	11.51	5.72	-80.91	-1946.01	-10176.6	-11.133	1.79	-1.981
1-chlorododecane	20	204.79	13.66	6.76	-94.58	-2257.66	-12762.2	-11.102	1.79	-1.981
1-chlorotetradecane	16	232.84	15.84	7.81	-108.28	-2569.33	-15471.7	-11.062	1.82	-1.981
1-chlorohexadecane	0	260.85	17.87	8.87	-121.98	-2881.02	-18288.4	-11.027	1.82	-1.981
1-chlorooctadecane	0	288.95	25.05	9.93	-135.53	-3192.67	-21210.1	-10.999	1.82	-1.981
1-bromoethane	92	108.97	2.71	1.61	-13.12	-678.77	-1658.6	-10.693	1.66	-1.994
1-bromobutane	108	137.03	5.18	2.75	-26.77	-990.44	-3381.1	-10.688	1.72	-1.997
1-bromohexane	78	165.08	7.56	3.82	-40.45	-1302.11	-5414.5	-10.688	1.75	-1.998
1-bromotetradecane	27	277.31	16.61	7.95	-95.24	-2548.79	-15405.2	-10.693	1.77	-1.998
1-iodobutane	73	184.02	5.39	3.05	-14.69	-984.41	-3348.8	-10.428	1.56	-1.991
1-iodopentane	65	198.05	5.98	3.58	-21.54	-1140.25	-4331.5	-10.428	1.58	-1.991
1-iodohexane	35	212.08	7.93	4.11	-28.38	-1296.08	-5379.6	-10.428	1.58	-1.991
1,1-dichloromethane	6	84.93	2.54	1.25	-25.85	-903.57	-1783.2	-11.391	1.51	-2.107
1,2-dichloroethane	13	98.96	7.99	1.48	-33.07	-1059.42	-2626.8	-11.387	2.23	-1.966
1,3-dichloropropane	152	112.99	5.12	2.01	-40.75	-1215.29	-3511.3	-11.372	1.51	-2.024
1,4-dichlorobutane	155	127.01	5.34	2.24	-48.09	-1371.15	-4497.1	-11.298	0	-2.004
1,6-dichlorohexane	113	155.07	7.69	3.29	-62.02	-1682.83	-6653.5	-11.222	0	-1.991
1,9-dichlorononane	66	197.15	11.99	4.88	-82.66	-2150.33	-10248.1	-11.179	1.54	-1.983
1,10-dichlorodecane	60	211.18	12.29	5.41	-89.53	-2306.17	-11523.9	-11.171	0.05	-1.982
1,2-dibromoethane	87	187.87	11.28	1.96	-7.56	-1018.35	-2522.4	-10.759	2.21	-1.985

[1] percentage of dehalogenation rate constant obtained with 1-chlorobutane [16]

rationally selected compounds resulted in several models with limited validity and applicability. The most robust model we have obtained is based on 24 compounds [16] (Table 4, Model 1). Some recent [22] and on-going experimental testing of additional compounds selected by experimental design should provide more data for structure-activity analysis.

Model 1.

Compounds domain: terminally substituted mono- and dihalogenated alkanes
Testing system: intact cells of *Rhodococcus erythropolis* Y2
Number of compounds: 24 + 3 outliers
Number of descriptors fixed in model: 9
Variance explained by first component: 78%
Descriptors important for the first component: logP, Mw, IX, EE, TE, Hf
Variance explained by second component: 14%
Descriptors important for the second component: HOMO, Dip, Hf, BCLU
Outliers: 1,1-dichloromethane; 1-chloropropane; 1,2-dichloroethane

Statistics: $R^2 = 0.92$; $Q^2 = 0.87$; n = 24

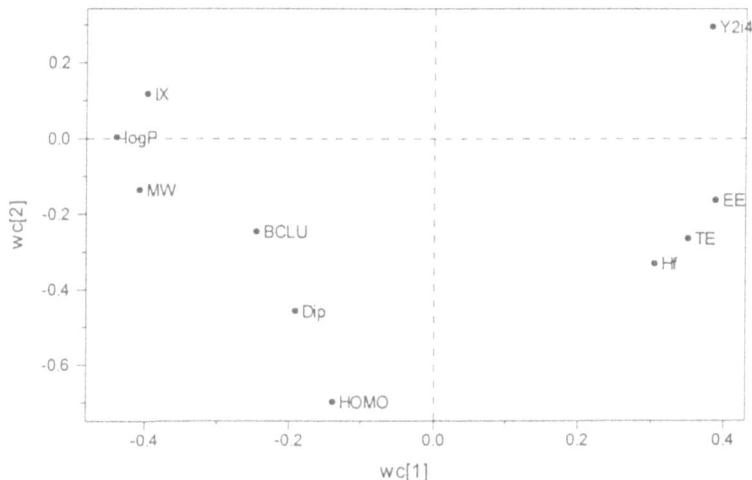

Figure 5.   Loading plot of the first two loading vectors obtained from PLS analysis.

Model interpretation: A group of 27 terminally halogenated compounds tested on 1-chlorobutane grown intact cells of *Rhodococcus erythropolis* Y2 [16] have been considered in PLS analysis which yielded a two component model. A large part of the variance modelled by the three most influential descriptors of the first component (logarithm of the octanol/water partition coefficient, logP, molecular weight, Mw, and moments of inertia along the X-axis, IX, Figure 5) accounts for the changes in size (length) and hydrophobicity of linear haloalkanes. Three other descriptors important for the first component (electronic energy EE, total energy, TE, and heat of formation, Hf, Figure 5) are electronic in nature. Since the penetration has been in previous discussion excluded as a rate limiting factor of a dehalogenation process, it is probable that this component describes the protein-ligand association which is dependent on steric and electronic complementarity between the ligand and enzyme active site, but can also correlate with hydrophobicity of haloalkanes with increasing chain-length. The last is especially true in case of haloalkane dehalogenases which have their active sites buried in the interior of the protein [36,40]. The second component is based exclusively on electronic descriptors (energy of the highest occupied molecular orbital, HOMO, dipole moment, Dip, heat of formation, Hf and bond contribution of the lowest unoccupied molecular orbital, BCLU) accounting mainly for the differences between chlorinated, brominated and iodinated compounds. It is unlikely that these descriptors relate (only) to protein-ligand association, but rather to the reaction mechanism of the dehalogenation reaction. This suggestion is consistent with our more recent results with modelling the dehalogenation of haloalkenes, where no long-chain compounds were present in data set (IX and logP were no longer significant) and electronic parameters were found to correspond with the proposed reaction mechanism.

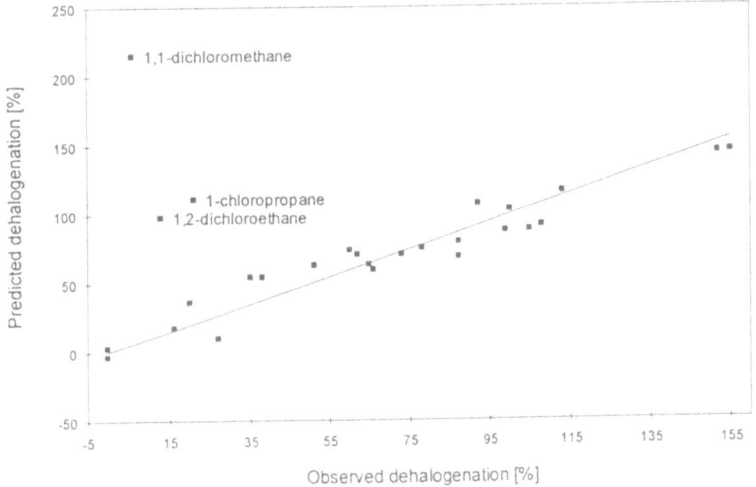

Figure 6.   Correlation plot of experimental versus predicted dehalogenation rates

Three outliers from the model have been detected: 1,1-dichloromethane, 1-chloropropane and 1,2-dichloroethane. All three are very short-chain compounds and displayed much lower dehalogenation than predicted from the model (Figure 6). Low dehalogenation rates for these compounds have been observed also in the assay with pure enzyme, excluding the possibility, that their recalcitrance is caused by toxicity of the dehalogenation intermediates - halogenated aldehydes - described for dehalogenation of 1,2-dibromoethane by *Xanthobacter autotrophicus* GJ10 [58]. It can be speculated, that the active site of enzyme Y2 haloalkane dehalogenase is 'optimized' to larger substrates and small molecules cannot adopt the correct orientation to be efficiently catalysed. The importance of the proper orientation of the substrate in the enzyme active site for biochemical catalysis has been described for example by Paulsen and Ornstein [59].

5. Conclusions

The application of a mechanistic approach for the study of mechanisms of microbial degradation processes and the development of Quantitative Structure-Biodegradability models is outlined in this contribution. Dehalogenation of haloaliphatic compounds was used as a case study and an attempt was made: (i) to determine the rate-limiting sub-process, (ii) to quantitatively estimate interspecies variability in substrate specificity and (iii) to investigate structure-activity relationships leading towards development of QSBR models.

A comparison of dehalogenation rates obtained in testing systems at different organization levels, intact cells and isolated enzyme, revealed that penetration of the halogenated aliphatic compounds into the cells of dehalogenase competent microbes is not the rate limiting step in hydrolytic dehalogenation. Consequently, the enzymatic reaction is to be considered slower than the penetration process. Further study of the reaction kinetics is needed to identify the rate-limiting step of the overall dehalogenation process. The applicability of molecular modelling techniques for such a study is illustrated.

Multivariate analysis of the substrate profiles of the haloalkane dehalogenases of different microbial strains was performed to tackle the problem of the utility of single species testing for prediction of environmental rate constants. At least two, probably four, separate groups of haloalkane dehalogenases have been formulated based on their ability to dehalogenate various haloaliphatic compounds. The QSBR model based on the data obtained from testing with the organism containing the dehalogenase from one group, cannot be safely used for predictions of biotransformation rates for organisms with dehalogenase of the other groups. Taking into account that the enzymes that act under the same reaction mechanism (hydrolytic dehalogenation) have been compared in this study, while at least seven different enzyme-catalysed dehalogenation reaction mechanisms are known [60,61] and there are, of course, many reactions besides dehalogenation which are important for decomposition of xenobiotic compounds, this result questions the utility of single-species derived QSBR models for the prediction of environmental rate constants. Nevertheless, we can imagine the situation when data from single-species test may well correspond to the rate constants observed in the environment. This may appear when tested compounds (training set) are structurally similar and the rate-limiting reaction of their decomposition is catalysed by a single enzyme which is present in the testing organism while at the same time widely spread among the natural degraders or when the penetration of the compounds into the cells is the rate-limiting step. The QSBR model based on data obtained from the single species test than corresponds to the model developed on data from mixed culture - environmental - tests [62] and can be used interchangeably for predictions.

A preliminary QSBR model for terminally substituted mono- and dihalogenated alkanes was developed. All three types of descriptors considered: hydrophobicity (logP), steric (Mw, IX), and electronic (EE, TE, Hf, HOMO, Dip, Hf, BCLU) were necessary to obtain a good description of the process studied. Three outliers from the model, all short-chain chlorinated compounds, were detected and the reasons for discrepancy between predictions and observation were discussed. It has been suggested, that the overall rate constant is controlled by an enzyme-substrate association (for long-chain compounds) and by chemical reaction. The model developed is based on data within a relatively small range of reaction rate constants and this fact needs to be taken into account in its possible future applications for predictive purposes. Although, more research is needed to better understand structure-biodegradability relationships for hydrolytic dehalogenation, the mechanistic approach has already provided beneficial information for model interpretation.

6. Acknowledgements

Haloalkane dehalogenase of R. *erythropolis* Y2 has been purified during the research internship of J.D. at the University of Kent at Canterbury (UK) sponsored by UNESCO within MIRCEN Biotechnology Fellowships. J.D. thanks to Dr. David Hardman and Prof. Alan Bull for supervising his work and to Dr. Helena M.S. Assis for help with the enzyme purification (all from University of Kent at Canterbury). Computationally extensive calculations have been performed on the computing facilities of the Czech Academic Supercomputer Centre in Brno and Prague. This work has been carried out under the framework of the project "Quantitative Structure-Activity Relationships for Predicting Fate and Effects of Chemicals in the Environment" under contract number EV5V-CT92-0211. Additional funding has been obtained in the framework of the EU Programme on Science and Technology Co-operation with Central and Eastern European countries under the supplementary agreement number CIPD-CT93-0042. Financial support from the European Union is gratefully acknowledged. Mary Lynam is acknowledged for kind help with the linguistic revision of the manuscript.

7. References

1. Boethling, R.S., and Sabljic, A. (1989) Screening-Level Model for Aerobic Biodegradability Based on a Survey of Expert Knowledge, *Environ. Sci. Technol.* **23**, 672-679.
2. Parsons, J.R., and Govers, H.A.J. (1990) Quantitative Structure-Activity Relationships for biodegradation, *Ecotox. Environ. Saf.* **19**, 212-227.
3. Degner, P., Muller, M., Nendza, M., and Klein, W. (1993) Structure-activity relationships for biodegradation, OECD Environment monographs, No. 68.
4. Peijnenburg, W. (1994) Structure-activity relationships for biodegradation: a Critical Review, *Pure Appl. Chem.* **66**, 1931-1941.
5. Alexander, M. (1981) Biodegradation of chemicals of environmental concern, *Science* **211**, 132-138.
6. Leisinger, T. (1983) Microorganisms and xenobiotic compounds, *Experientia* **39**, 1183-1191.
7. Gibson, D.T. (1993) Biodegradation, biotransformation and the Belmont, *J. Industr. Microbiol.* **12**, 1-12.
8. Providenti, M.A., Lee, H., and Trevors, J.T. (1993) Selected factors limiting the microbial degradation of recalcitrant compounds, *J. Indust. Microbiol.* **12**, 379-395.

90

9. Singleton, I. (1994) Microbial metabolism of xenobiotics: fundamental and applied research, *J. Chem. Tech. Biotechnol.* **59**, 9-23.

10. Damborský, J. (1996) A Mechanistic approach to deriving Quantitative Structure-Activity
    - Relationship models for microbial degradation of organic compounds, *SAR QSAR Environ. Res.* **5**, 27-36.

11. Sallis, P.J., Armfield, S.J., Bull, A.T., and Hardman, D.J. (1990) Isolation and characterization of a haloalkane halidohydrolase from *Rhodococcus erythropolis* Y2, *J. Gen. Microbiol.* **136**, 115-120.

12. Keuning, S., Janssen, D.B., and Witholt, B. (1985) Purification and characterization of hydrolytic haloalkane dehalogenase from *Xanthobacter autotrophicus* GJ10, *J. Bacteriol.* **163**, 635-639.

13. Scholtz, R., Leisinger, T., Suter, F., and Cook, A.M. (1987) Characterization of 1-Chlorohexane halidohydrolase, a dehalogenase of wide substrate range from an *Arthrobacter* sp., *J. Bacteriol.* **169**, 5016-5021.

14. Yokota, T., Omori, T., and Kodama, T. (1987) Purification and properties of haloalkane dehalogenase from *Corynebacterium sp.* Strain m15-3, *J. Bacteriol.* **169**, 4049-4054.

15. Janssen, D.B., Gerritse, J., Brackman, J., Kalk, C., Jager, D., and Witholt, B. (1988) Purification and characterization of a bacterial dehalogenase with activity toward halogenated alkanes, alcohols and ethers, *Eur. J. Biochem.* **171**, 67-92.

16. Armfield, S.J. (1990) Microbial dehalogenation of halogenated aliphatic compounds. PhD. Thesis, University of Kent at Canterbury.

17. Vienravi, V., 1993, Evolution of Haloalkane Dehalogenase in *Pseudomonas* sp. E4M. Ph.D. Thesis, University of Kent at Canterbury.

18. Nyandoroh, M.G., Hardman, D.J., and Bull, A.T. (1995) personal communication.

19. Nagata, Y., Miyauchi, K., Damborský, J., Manová, K., Ansorgová, A., and Takagi, M. (1996) Purification and characterization of haloalkane dehalogenase (LinB) from *Sphingomonas paucimobilis*, in preparation.

20. Janssen, D.B., Scheper, A., and Witholt, B. (1984) Biodegradation of 2-chloroethanol and 1,2-dichloroethane by pure bacterial cultures, in E.H. Houwink and R.R. van der Meer (eds.), *Innovations in Biotechnology*, Elsevier Science Publishers, Amsterdam, pp. 169-178.

21. Iwasaki, I., Utsumi, S., and Ozawa, T. (1952) New colorimetric determination of chloride using mercuric thiocyanate and ferric ion, *Bull. Chem. Soc. Japan* **25**, 226.

22. Damborský, J., Manová, K., Berglund, A., Sjöström, M., Němec, M., and Holoubek, I. (1996) Biotransformation of chloro- and bromoalkenes by crude extracts of *Rhodococcus erythropolis* Y2, submitted.

23. Leo, D., and Weininger, D. (1989) MedChem software manual, DayLight Chemical Information Systems, Inc., Irwine, CA.

24. OM (1993) TSAR v2.2, Oxford Molecular, Ltd., Oxford, UK.

25. Biosym/MSI (1995) InsightII, Biosym Technologies, San Diego, CA.

26. Stewart, J.J.P. (1990) MOPAC manual v 6.0, Quantum Chemistry Program Exchange.

27. Wold, S., Albano, C., Dunn III, W.J., Edlund, U., Esbensen, K., Geladi, P., Hellberg, S., Johansson, E., Lindberg, W., and Sjöström, M. (1984) Multivariate data analysis in chemistry, in B.R. Kowalski (ed.), *Chemometrics*, D. Reidel Publishing Company, Dordrecht, pp. 17-95.

28. UMETRI (1994) SIMCA-S 5.1b, Umetri AB, Umea, S.

29. Wold, S. (1991) Validation of QSAR's, *Quant. Struc.-Act. Relat.* **10**, 191-193.

30. Kato, Y., Itai, A., and Iitaka, Y. (1987) A novel method for superimposing molecules and receptor mapping, *Tetrahedron* **43**, 5229-5236.

31. Ho, C.M.W., and Marshall, G.R. (1990) Cavity search: an algorithm for the isolation and display of cavity-like binding regions, *J. Comput.-Aid. Mol. Design* **4**, 337-354.

32. Cosentino, U., Moro, G., Pitea, D., Scolastico, S., Todeschini, R., and Scolastico, C. (1992) Pharmacophore identification by molecular modeling and chemometrics: The case of HMG-CoA reductase inhibitors, *J. Comput.-Aid. Mol. Design* **6**, 47-60.

33. Gouldson, P.R., Winn, P.J., and Reynolds, C.A. (1995) A molecular dynamics approach to receptor mapping: Application to the 5HT(3) and beta(2)-adrenergic receptors, *J. Med. Chem.* **38**, 4080-4086.

34. Eisenmenger, F., Argos, P., and Abagyan, R. (1993) A method to configure protein side-chains from the main-chain trace in homology modelling, *J. Mol. Biol.* **231**, 849-860.

35. Snow, M.E. (1993) Computational studies: protein homology modelling, *Chem. Design. Automat. News* **8**, 11-14.

36. Damborský, J., Bull, A.T., and Hardman, D.J. (1995) Homology modelling of the haloalkane dehalogenase of *Sphingomonas paucimobilis* UT26, *Biologia* **50**, 523-528.

37. Wibley, J.E.A., McKie, J.H., Embrey, K., Marks, D.S., Douglas, K.T., Moore, M.H., and Moody, P.C.E. (1995) A homology model of the three-dimensional structure of human O6-Alkylguanine-DNA Alkyltransferase based on the crystal structure of the C-terminal domain of the Ada protein from *Escherichia coli, Anti-Canc. Drug Design* **10**, 75-95.

38. Poulos, T.L., Finzel, B.C., and Howard, A.J. (1986) Crystal structure of substrate-free *Pseudomonas putida* cytochrome P-450, *Biochemistry* **25**, 5314-5322.

39. Ohlendorf, D.H., Lipscomb, J.D., and Weber, P.C. (1988) Structure and assembly of protocatechuate 3,4-dioxygenase, *Nature* **336**, 403-405.

40. Rozeboom, H.J., Kingma, J., Janssen, D.B., and Dijkstra, B.W. (1988) Crystallization of haloalkane hehalogenase from *Xanthobacter autotrophicus* GJ10, *J. Mol. Biol.* **200**, 611-612.

41. Rosenzweig, A.C., Fredrick, C.A., Lippard, S.J., and Norlund, P. (1993) Crystal Structure of Bacterial Non-haeme Iron Hydroxylase that Catalyses the Biological Oxidation of Methane, *Nature* **366**, 537-543.

42. Han, S., Eltis, L.D., Timmis, K.N., Muchmore, S.W., and Bolin, J.T. (1995) Crystal structure of the biphenyl-cleaving extradiol dioxygenase from a PCB-degrading Pseudomonad, *Science* **270**, 976-980.

43. Sugiyama, K., Narita, H., Yamamoto, T., Senda, T., Kimbara, K., Inokuchi, N., Iwama, M., Irie, M., Fukuda, M., Yano, K., and Mitsui, Y. (1995) Crystallization and preliminary crystallographic analysis of a 2,3-dihydroxybiphenyl dioxygenase from *Pseudomonas* sp strain KKS102 having polychlorinated biphenyl (PCB)-degrading activity, *Proteins - Struc. Funct. Gen.* **22**, 284-286.

44. Ridder, I.S., Rozeboom, H.J., Kingma, J., Janssen, D.B., and Dijkstra, B.W. (1995) Crystallization and preliminary X-ray analysis of L-2-haloacid dehalogenase from *Xanthobacter autotrophicus* GJ10, *Prot. Sci.* **4**, 2619-2620.

45. Verschueren, K.H.G., Franken, S.M., Rozeboom, H.J., Kalk, K.H., and Dijkstra, B.W. (1993) Refined x-ray structures of haloalkane dehalogenase at pH 6.2 and pH 8.2 and implications for the reaction mechanism, *J. Mol. Biol.* **232**, 856-872.

46. Verschueren, K.H.G., Franken, S.M., Rozeboom, H.J., Kalk, K.H., and Dijkstra, B.W. (1993) Non-covalent binding of the heavy atom compound [Au(CN)2]- at the halide binding site of haloalkane dehalogenase from *Xanthobacter autotrophicus* GJ10, *FEBS Lett.* **323**, 267-270.

47. Verschueren, K.H.G., Kingma, J., Rozeboom, H.J., Kalk, K.H., Janssen, D.B., and Dijkstra, B.W. (1993) Crystallographic and fluorescence studies of the interaction of haloalkane dehalogenase with halide ions. Studies with halide compounds reveal a halide binding site in the active site, *Biochemistry* **32**, 9031-9037.

48. Verschueren, K.H.G., Seljee, F., Rozeboom, H.J., Kalk, K.H., and Dijkstra, B.W. (1993) Crystallographic analysis of the catalytic mechanism of haloalkane dehalogenase, *Nature* **363**, 693-698.

49. Damborský, J., Kutý, M., Němec, M., and Koča, J. (1996) Molecular modelling study of the catalytic mechanism of haloalkane dehalogenase: I. quantum chemical study of the first reaction step., submitted.

50. Bash, P.A., Field, M.J., and Karplus, M. (1987) Free energy perturbation method for chemical reactions in the condensed phase: a dynamical approach based on a combined quantum and molecular mechanics potential, *J. Am. Chem. Soc.* **109**, 8092-8094.

51. Bash, P.A., Field, M.J., Davenport, R.C., Petsko, G.A., Ringe, D., and Karplus, M. (1991) Computer simulation and analysis of the reaction pathway of triosephosphate isomerase, *Biochemistry* **30**, 5826-5832.

52. Rorije, E., Richter, J., and Peijnenburg, W.J.G.M. (1994) Modeling anaerobic reductive dehalogenation using MOPAC-calculated activation energies, Conference *Quantitative Structure-Activity Relationships (QSAR) in Environmental Sciences*, p. 37. Belgirate, Italy.

53. Hermens, J., Balaz, S., Damborský, J., Karcher, W., Müller, M., Peijnenburg, W., Sabljić, A., and Sjöström, M. (1995) Assessment of QSARs for predicting fate and effects of chemicals in the environment: an international european project, *SAR QSAR Environ. Res.* **3**, 223-236.

54. Damborský, J., Nyandoroh, M.G., Nemec, M., Holoubek, I., Bull, A.T. and Hardman, D.J. (1996) Haloalkane dehalogenases: a comparative study, submitted.

55. Yokota, T., Fuse, H., Omori, T., and Minoda, Y. (1986) Microbial dehalogenation of halo-alkanes mediated by oxygenase or halidohydrolase, *Agric. Biol. Chem.* **50**, 453-460.

56. Eriksson, L., Jonsson, J., and Tysklind, M. (1995) Multivariate QSBR modeling of biodehalogenation half-lives of halogenated aliphatic hydrocarbons, *Environ. Toxicol. Chem.* **14**, 209-217.

57. Franken, S.M., Rozeboom, H.J., Kalk, K.H., and Dijkstra, B.W. (1991) Crystal structure of haloalkane dehalogenase: an enzyme to detoxify halogenated alkanes, *EMBO J.* **10**, 1297-1302.

58. Pries, F., Van der Ploeg, J.R., Wijngaard, A.J., Bos, R., and Janssen, D.B., 1994, Adaptation of bacteria to chlorinated hydrocarbon degradation, in R.E. Hinchee, A. Leeson, L. Semprini, and S.K. Ong (eds.), *Bioremediation of Chlorinated and Polycyclic Aromatic Hydrocarbon Compounds*, Lewis Publishers, Boca Raton, pp. 259-265.

59. Paulsen, M.D., and Ornstein, R.L. (1994) Active-site mobility inhibits reductive dehalogenation of 1,1,1-trichloroethane by cytochrome P450cam, *J. Comput.-Aid. Mol. Design* **8**, 389-404.

60. Fetzner, S., and Lingens, F. (1994) Bacterial dehalogenases: biochemistry, genetics, and biotechnological applications, *Microbiol. Rev.* **58**, 641-685.

61. Janssen, D.B., Pries, F., and VanderPloeg, J.R. (1994) Genetics and biochemistry of dehalogenating enzymes, *Annu. Rev. Microbiol.* **48**, 163-191.

62. Damborský, J., Eriksson, L., and Hermens, J.L.M. (1993) QSAR for microbial transformation of phenols, Conference TOCOEN 93', Znojmo, Czech Republic.

QUANTITATIVE STRUCTURE-BIODEGRADABILITY STUDIES: AN INVESTIGATION OF THE MITI AROMATIC COMPOUND DATA-BASE

J.C. DEARDEN and M.T.D. CRONIN
School of Pharmacy and Chemistry,
Liverpool John Moores University
Byrom Street, Liverpool L3 3AF, UNITED KINGDOM

1. Introduction

Risk is a function of both intrinsic hazard and exposure. Knowledge of the persistence of compounds in the environment is thus of prime importance in the assessment of risk; clearly, therefore, ability to predict persistence (or lack of persistence, i.e. biodegradability) can be of great help in this respect.

Some years ago we obtained a 240 compound data-set of biodegradability data, in the form of 5-day BOD (biological oxygen demand) values (as % ultimate degradation), measured using acclimated sludge; the data were supplied by the U.S. Environmental Protection Agency Laboratory in Duluth, MN. They had not been determined there, but had been carefully screened for consistency of determination.

We broke the data down into chemical classes, and searched for quantitative structure-biodegradability relationships (QSBRs) with a range of relevant physico-chemical and structural parameters. One parameter that appeared to be significant was the modulus of the difference of atomic charge across one of the bonds in a molecule; this parameter was selected because of its potential to model the strength of a bond. For example, we found [1] the following QSBR for a series of 11 phenols:

$$BOD = 0.998 \times 10^3 \, \Delta|\delta \, |_{C-O} + 2.108 \tag{1}$$

$$n = 11 \quad r = 0.991 \quad s = 4.04$$

However, it should be pointed out that the CINMIN program used to calculate atomic charges was not a standard program, and it has not been possible to reproduce these results using programs such as MOPAC. This must cast some doubt on the universal applicability of the results. We remain convinced, however, that a parameter that reflects bond strength should be able to contribute to the modelling of biodegradability.

We also found [1] that 5-day BOD values correlated well with electrophilic super-delocalisability (S_E). For example, for alcohols we found a good correlation with S_E on the carbon to which the hydroxyl group was attached:

93

W. J. G. M. Peijnenburg and J. Damborský (eds.), Biodegradability Prediction, 93–104.
© 1996 *Kluwer Academic Publishers.*

$$BOD = 0.093 \, S_E - 3.163 \qquad (2)$$

$$n = 19 \quad r = 0.981 \quad s = 4.30$$

The superdelocalisabilities were calculated for energy-minimised molecules using MOPAC.

We have also carried out work [2] on the correlation of biodegradability with molecular connectivity indices, as have Boethling and co-workers [3,4]. For a series of 20 alcohols, we found a good correlation with the difference of third-order simple and valence-corrected molecular connectivities:

$$BOD = 12.40 \, (^3\chi - {}^3\chi^v) + 0.52 \qquad (3)$$

$$n = 20 \quad r = 0.952 \quad s = 7.67$$

Clearly the size and shape of molecules can play an important role in their bio-degradability, and several workers have reported QSBRs involving steric parameters [5]. We observed [6] a correlation between BOD values and the Sterimol substituent length L for halogenated hydrocarbons:

$$BOD = 8.29 \, L - 1.187 \qquad (4)$$

$$n = 9 \quad r = 0.976 \quad s = 4.12$$

We further observed a good correlation of BOD values of alkanes with accessible surface area (ASA):

$$BOD = 0.0996 \, ASA + 0.055 \qquad (5)$$

$$n = 12 \quad r = 1.000 \quad s = 0.27$$

These parameters did not correlate well with the biodegradability of polar compounds, and indeed such correlations would not be expected.

More recent work from our laboratory [7] has investigated the application of Comparative Molecular Field Analysis (CoMFA) to the correlation of biodegradability. This technique investigates the steric and electrostatic fields sensed by a probe at different points within a three-dimensional grid within which the molecules are aligned. No multiple linear regression-type QSAR equations are produced, but correlation coefficients indicate the degree of fit. We examined a number of different data-sets, and reasonable to good fits were obtained for all except a series of phenols. For example, for a series of 20 alcohols, an r^2 of 0.980 (and cross-validated (CV) r^2 of 0.724) was obtained, with 74.7 % steric and 25.3 % electrostatic contribution to biodegradation. For 20 linear alkylbenzene sulphonates, $r^2 = 0.913$ and $CVr^2 = 0.761$, with the degradation being 100 % controlled by steric factors. This technique is generally applicable only to congeneric series, otherwise the alignment problems are very difficult.

2. The MITI Aromatic Compound Biodegradation Data-Base

We have recently obtained, courtesy of Dr. W.J.G.M. Peijnenburg, the biodegradation data for 240 aromatic compounds determined by MITI. The data-set comprised mostly compounds with a single aromatic ring, but there were 17 compounds with two aromatic rings and 10 compounds with three aromatic rings. There were no fused-ring compounds in the data-set, with the exception of two anhydrides. Substitution patterns were extremely varied, with numbers of substituents up to six per ring: there was a considerable number of chlorine-containing compounds. Molecular weights ranged from 78 (benzene) to 959 (decabromodiphenyl ether). Four compounds were sodium salts; these were treated as the corresponding acids. A few compounds were indicated as being mixtures of isomers; in such cases the most appropriate isomer was used. Three compounds could not be handled by the software that we used; they contained respectively iodine, tin and silicon.

For each compound, the data supplied were: ID number, CAS number, molecular structure, molecular formula, molecular weight, % BOD (biological oxygen demand) at 7 days, % BOD at 14 days, induction time to 5 % BOD, % BOD at 10 days after 5 % BOD, maximum biodegradation rate, test period, and % BOD at the end of the test. 122 of the compounds showed some biodegradation, whilst 138 did not (i.e. % BOD at end of test was zero). In a few cases replicate results were given; in such cases the first result was used, for consistency with the other data.

2.1 METHODS

For each compound a large number of descriptors was calculated. Log P values were calculated using CLOGP for Windows; for the few compounds for which CLOGP was unable to calculate a value because of missing fragments, values were calculated using ProLogP or, in one case, MicroQSAR. Molar refractivities were calculated using MEDCHEM ver. 3.55. For the calculation of molecular orbital properties, molecules were constructed and energy-minimised in COSMIC, and then MOPAC ver. 7.0 was used to calculate heat of formation, E_{HOMO} (energy of the highest occupied molecular orbital), E_{LUMO} (energy of the lowest unoccupied molecular orbital) and dipole moment. MOLCONN-X was used to calculate a large number of topological indices, including molecular connectivities up to tenth order, electrotopological indices and information contents. The method of Fujita et al. [8] was used to determine the hydrogen bond donor and acceptor abilities of each molecule, and the method of Yang et al. [9] to determine the number of donor and acceptor hydrogen bonds that each molecule was capable of forming. Statistical analysis was carried out using MINITAB ver. 10.1.

2.2 RESULTS AND DISCUSSION

Virtually all of the QSAR work carried out to date on biodegradability has been done on congeneric series of compounds, in which mechanisms of biodegradation might be expected to be reasonably consistent. The present data set is extremely diverse, and *a priori* one would not expect to find good QSAR correlations. This was indeed found to be the case with the MITI data-set. There were 79 compounds for which all (or most) of the measured biodegradability end-points were given. Step-wise regression showed that the best correlation that could be obtained with final BOD values was:

$$BOD = -28.2 \, ^3\chi^v - 0.00193 \, I_{Bon} - 0.184 \, HeatForm + 71.0 \qquad (6)$$

$$n = 79 \quad r^2(adj.) = 0.180 \quad s = 23.3 \quad F = 6.72$$

where $^3\chi^v$ = third order valence-corrected path molecular connectivity, I_{Bon} = Bonchev index, and HeatForm = heat of formation.

It is, of course, probable that considerably better correlations could be obtained if single chemical classes were examined, but we were concerned here to determine whether the data would allow of a reasonable prediction of biodegradability across a wide range of chemical classes, using a single regression equation. Clearly that was not the case. Similarly poor correlations were obtained with all the other biodegradability end-points: 7-day BOD, r^2 = 0.122; 14-day BOD, r^2 = 0.255; time to 5 % biodegradation, r^2 = 0.232; % BOD at 10 days after 5 % BOD, r^2 = 0.153; maximum biodegradation rate, r^2 = 0.229.

It may be noted here that in general there is not a particularly high correlation between the biodegradability end-points, as Table 1 shows. The high correlation between 14-day BOD and final BOD is because many of the compounds had achieved their final BOD by 14 days.

TABLE 1. Correlation matrix (r) between biodegradability end-points

	7-d BOD	14-d BOD	T5 %	10-day	Rmax
14-d BOD	0.771				
T5 %	-0.751	-0.719			
10-day	0.792	0.873	-0.618		
Rmax	0.546	0.497	-0.459	0.491	
Final BOD	0.634	0.878	-0.610	0.868	0.372

At this point in our analysis we observed a number of errors in the MITI data-base. One example will serve to illustrate these; for 3-nitrophenol, the 7-day BOD (63 %) and the 14-day BOD (65 %) values were both considerably higher than the final BOD value of 46 %. It may also be noted that in some instances where replicate BOD values were given for a compound, the replicate value differed considerably from the original determination; for example, for 3-nitrobenzoic acid the two reported final BOD values were 0 % and 12 %.

In view of this, and because of the lack of quantitative correlation of the biodegradability end-points, we decided to investigate whether any of the parameters that we had generated could distinguish between readily and non-readily biodegradable compounds. The former we defined as those compounds with a final BOD greater than 50 %, and the latter as those with a final BOD of less than 20 %. The number of compounds with final BOD values between 20 and 50 % was only 15, although one compound (CAS no. 60628-17-3) had replicate final BOD values of 14 and 25 %; we included it as a non-readily biodegradable compound. Step-wise regression indicated that three parameters, namely heat of formation, sixth order chain molecular connectivity and the total topological index based on electrotopological state indices (the last two being obtained from MOLCONN-X), were important in modelling final BOD. Discriminant analysis on the 222 compounds, using these three parameters, yielded 172 correct predictions, i.e. a 77.5 % correct prediction rate. Of these, there were 117/160 (73.1 %) correct predictions for non-readily biodegradable compounds, and 55/62 (88.7 %) correct for readily biodegradable compounds. Whilst these prediction rates are reasonable, they nevertheless leave something to be desired, and so we examined which groups of compounds were outliers. Phosphorus-containing compounds were the most obvious in this respect, and a highly substituted fluorine compound (CAS no. 346-10-1) was observed to have an extremely influential effect; these compounds (which included that mentioned earlier with CAS no. 60628-17-3) were therefore removed from the data-set to give some improvement in the prediction rate. For the 210 compounds remaining, there were 114/150 (76.0 %)correct predictions for non-readily biodegradable compounds, and 53/60 (88.3 %) correct predictions for readily biodegradable compounds; the overall correct prediction rate was therefore 167/210 (79.5 %). Cross-validation yielded 113/150 (75.3 %) correct predictions for non-readily biodegradable compounds and 52/60 (86.7 %) correct for readily biodegradable compounds, with an overall correct prediction rate of 165/210 (78.6 %). Figure 1 shows the discrimination observed for the 210 compound data-set.

The heat of formation of a compound is a measure of its stability, and hence would be expected to reflect the ability of the compound to biodegrade. TETS2 is a topological index which is the sum of all path terms computed with E-state values [10]; it contains both electronic and topological information, but, being a newly-derived index, has not yet been fully evaluated. It is believed to relate to, *inter alia*, the compactness of a molecule. Sixth order chain molecular connectivity represents essentially the number of substituents on an aromatic ring, with high values of ChainS6 indicating fewer substituents. Figure 1 clearly shows that readily biodegradable compounds generally have high ChainS6 values, and one would expect that the more substituents on an aromatic ring, the less readily should the compound biodegrade. It is pleasing that as few as three readily calculated parameters are able to give good discrimination between readily and non-readily biodegradable aromatic compounds with widely diverse substituents.

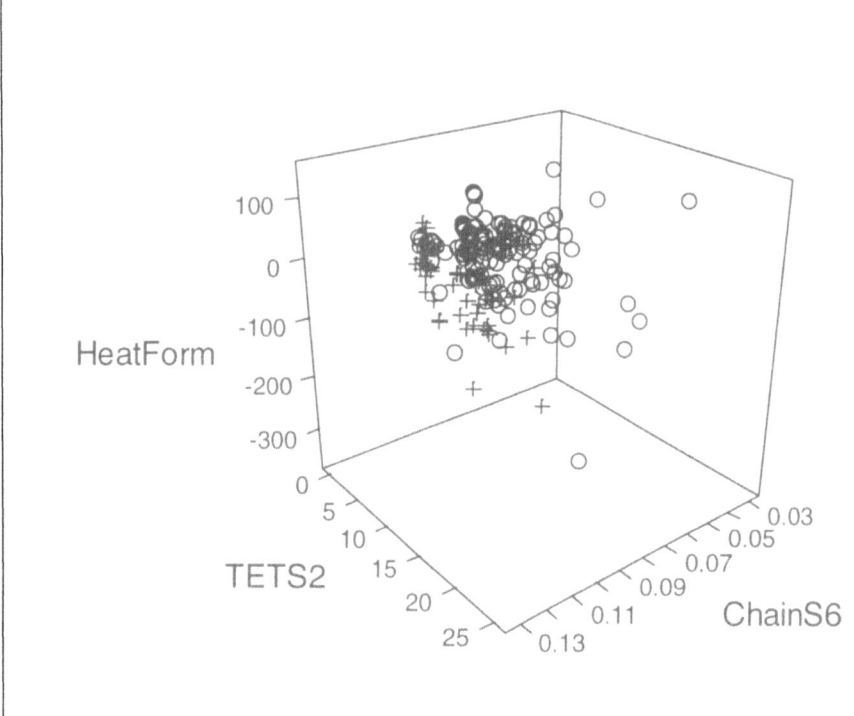

Figure 1. Discrimination between readily (+) and non-readily (o) biodegradable compounds (HeatForm = heat of formation, TETS2 = total topological index based on electrotopological state indices, ChainS6 = sixth order chain molecular connectivity)

Table 2. CAS numbers, final BOD values and physico-chemical properties of compounds used in this analysis

CAS number	Final BOD	Heat of Formation (kcal mol^{-1})	Chain 6	TETS2
71-43-2	40.0	22.028	0.1250	2.8967
108-88-3	100.0	14.423	0.1021	3.7304
100-42-5	53.0	38.719	0.1021	4.4407
99-87-6	88.0	-2.600	0.0833	6.0944
98-51-1	52.0	-3.978	0.0833	6.5193
100-41-4	39.0	8.678	0.1021	4.4299
98-87-3	90.0	3.182	0.1021	4.5682
100-44-7	71.0	6.919	0.1021	4.0774
103-84-4	69.0	-13.785	0.1021	4.4210
104-42-7	62.0	-56.560	0.0833	10.2502
95-53-4	34.0	17.713	0.0833	4.2441

2415-85-2	74.0	-61.046	0.0833	6.1738
64-04-0	90.0	13.499	0.1021	4.7743
100-46-9	64.0	19.557	0.1021	4.1024
588-46-5	77.0	-20.976	0.1021	4.6712
108-95-2	85.0	-22.237	0.1021	2.8405
1319-77-3	50.0	-28.853	0.0833	3.5102
1300-71-6	44.0	-36.366	0.0680	4.3271
108-46-3	67.0	-66.582	0.0833	1.9427
120-80-9	83.0	-64.368	0.0833	1.9070
123-31-9	70.0	-65.678	0.0833	2.3202
100-66-3	56.0	-15.814	0.1021	4.4085
103-71-1	63.0	-21.290	0.1021	5.0795
122-60-1	24.0	-18.849	0.1021	6.8248
90-05-1	90.0	-59.913	0.0833	3.6106
150-76-5	86.0	-59.195	0.0833	4.0848
91-16-7	96.0	-50.196	0.0833	5.8127
93-15-2	89.0	-37.736	0.0680	8.2197
614-80-2	60.0	-58.074	0.0833	3.8270
90-04-0	55.0	-16.254	0.0833	4.7196
104-94-9	65.0	-16.506	0.0833	4.9145
554-84-7	46.0	-17.420	0.0833	2.8290
25167-82-2	83.0	-38.288	0.0556	2.5716
118-79-6	49.0	-2.181	0.0556	4.3044
100-51-6	94.0	-28.248	0.1021	3.4687
140-11-4	95.0	-67.595	0.1021	5.3352
60-12-8	87.0	-35.265	0.1021	4.2937
96-09-3	81.0	18.269	0.1021	5.8385
104-51-1	91.0	-15.034	0.1021	5.1292
536-60-7	85.0	-47.736	0.0833	5.6787
100-52-7	66.0	-8.916	0.1021	3.7031
98-86-2	67.0	-14.988	0.1021	3.7503
118-90-1	94.0	-69.521	0.0833	4.8199
131-11-3	93.0	-137.333	0.0833	5.9474
84-61-7	69.0	-178.537	0.1021	10.9007
85-68-7	81.0	-110.660	0.0833	12.2276
1459-93-4	99.0	-128.823	0.0833	6.3923
131-17-9	82.0	-83.961	0.0833	6.8795
120-61-6	84.0	-143.006	0.0833	6.5230
121-91-5	78.0	-156.798	0.0833	3.2081
100-21-0	75.0	-144.634	0.0833	4.2001
536-66-3	89.0	-85.482	0.0833	6.1882
85-44-9	85.0	-68.834	0.1021	9.1486
94-36-0	84.0	-31.435	0.1021	11.9686
552-30-7	96.0	-148.639	0.0833	9.1099
1726-23-4	88.0	-277.264	0.0680	9.5852
495-69-2	81.0	-100.676	0.1021	6.1007
65-85-0	85.0	-61.976	0.1021	4.1355
119-68-6	85.0	-59.805	0.0833	5.3683
62-23-7	62.0	-55.332	0.0833	3.7476
552-16-9	100.0	-57.503	0.0833	5.7653
87-20-7	83.0	-118.109	0.0833	4.9636

69-72-7	88.0	-106.842	0.0833	3.2946
100-47-0	63.0	53.383	0.1021	3.9994
140-29-4	87.0	46.694	0.1021	4.5141
98-11-3	87.0	-97.075	0.1021	6.1908
1300-51-2#	86.0	-139.089	0.0833	6.3828
115-86-6*	90.0	-140.218	0.1021	27.3580
15205-57-9*	79.0	-104.387	0.1021	22.1709
1330-78-5*+	28.0	-158.872	0.0833	31.5726
1330-78-5*+	37.0	-163.251	0.0833	31.4299
122-40-7	51.0	-25.253	0.1021	8.2887
618-41-7#	22.0	-27.515	0.1021	4.7563
98-83-9	0.0	32.463	0.1021	5.2439
95-63-6	18.0	-0.098	0.0680	5.6159
108-67-8	0.0	-0.677	0.0680	5.6244
526-73-8	0.0	1.220	0.0680	5.6144
28106-30-1	0.0	25.390	0.0833	6.1971
535-77-3	0.0	-2.834	0.0833	6.1004
105-05-5	0.0	-4.709	0.0833	6.2203
141-93-5	0.0	-4.661	0.0833	6.2331
1321-74-0	0.0	55.452	0.0833	6.1497
25321-09-9	0.0	-12.480	0.0833	7.7169
108-90-7	0.0	14.810	0.1021	3.4256
108-86-1	0.0	26.758	0.1021	3.6653
95-49-8	0.0	7.918	0.0833	4.2127
106-43-4	0.0	7.121	0.0833	4.2663
95-50-1	0.0	9.220	0.0833	3.6624
541-73-1	0.0	8.199	0.0833	3.8133
1712-70-5	0.0	25.247	0.0833	5.7634
106-37-6	0.0	32.026	0.0833	4.4633
98-15-7	0.0	-140.453	0.0833	8.1972
87-83-2	7.0	47.058	0.0370	7.6061
87-61-6	0.0	4.197	0.0680	3.6587
108-70-3	0.0	2.228	0.0680	4.0119
120-82-1	0.0	2.930	0.0680	3.9030
95-94-3	0.0	-1.874	0.0556	3.7376
118-74-1	0.0	-7.822	0.0370	1.7108
608-93-5	0.0	-4.940	0.0454	2.9044
634-66-2	0.0	-0.600	0.0556	3.4390
98-08-8	0.0	-134.366	0.1021	8.2994
102-47-6	0.0	-5.300	0.0680	4.6483
100-61-8	1.4	26.702	0.1021	4.5374
579-10-2	1.0	-1.076	0.1021	5.2858
91-66-7	0.0	21.541	0.1021	7.1427
121-69-7	1.9	32.363	0.1021	5.5464
103-69-5	0.0	17.994	0.1021	5.2343
95-80-7	0.0	15.398	0.0680	4.6623
95-64-7	7.1	6.129	0.0680	5.1915
95-78-3	1.0	10.178	0.0680	5.1386
122-39-4	0.0	58.488	0.1021	11.3147
26603-23-6	0.0	-43.856	0.0833	23.9307
74-31-7	0.2	91.657	0.1021	21.2089

108-45-2	2.0	19.135	0.0833	3.8791
108-44-1	1.0	16.851	0.0833	4.2850
611-21-2	1.0	20.706	0.0833	5.4894
622-57-1	2.0	10.427	0.0833	6.2443
696-44-6	0.0	16.825	0.0833	5.4977
95-51-2	2.7	13.129	0.0833	3.7024
95-76-1	0.0	11.800	0.0680	3.9434
95-82-9	0.0	6.288	0.0680	3.9542
554-00-7	0.0	6.293	0.0680	3.9512
95-69-2	0.0	5.843	0.0680	4.6650
634-93-5	0.0	0.072	0.0556	3.7195
1477-55-0	22.0	14.953	0.0833	5.3587
103-83-3	2.0	26.509	0.1021	5.9811
88-74-4	0.0	26.125	0.0833	2.8185
99-09-2	0.0	27.909	0.0833	2.9270
100-01-6	0.0	21.694	0.0833	3.2963
89-63-4	0.0	16.331	0.0680	3.4248
98-95-3	3.3	26.105	0.1021	3.2178
88-72-2	0.5	18.939	0.0833	3.8216
99-08-1	2.0	17.805	0.0833	3.8949
99-99-0	0.8	17.292	0.0833	3.9259
88-73-3	8.2	23.863	0.0833	2.8682
100-00-5	0.0	19.284	0.0833	3.2425
25154-54-5	0.0	33.141	0.0833	4.9541
610-39-9	0.0	30.538	0.0680	6.5450
118-83-2	0.0	-128.507	0.0680	12.8858
99-54-7	0.0	15.112	0.0680	3.0826
102-06-7	0.0	114.907	0.1021	11.6025
98-54-4	0.0	-40.457	0.0833	5.0281
99-71-8	0.0	-45.396	0.0833	5.4647
1806-26-4	0.0	-76.637	0.0833	7.5271
25154-52-3	0.0	-83.486	0.0833	7.9916
5510-99-6	3.0	-64.103	0.0680	8.0938
96-76-4	0.0	-56.296	0.0680	6.2135
128-37-0	4.5	-59.138	0.0556	6.4651
732-26-3	0.0	-69.606	0.0556	8.7923
4130-42-1	0.0	-63.397	0.0556	7.3681
101-53-1	0.0	-2.060	0.1021	10.4433
67557-76-0	0.0	-39.365	0.0833	10.4016
120-71-8	0.7	-24.084	0.0680	5.6133
101-84-8	6.3	23.593	0.1021	10.9806
591-27-5	0.0	-24.030	0.0833	3.0575
156-43-4	0.0	-22.076	0.0833	5.5998
62-44-2	8.4	-56.976	0.0833	6.8836
346-10-1*	1.0	-476.510	0.0680	59.0480
88-75-5	0.0	-20.163	0.0833	3.8454
100-02-7	4.3	-19.549	0.0833	1.9265
91-23-6	0.0	-9.585	0.0833	2.8586
100-17-4	0.0	-13.449	0.0833	4.1996
555-03-3	1.0	-11.381	0.0833	3.8770
119-33-5	0.0	-27.609	0.0680	4.3638

2042-14-0	0.0	-23.572	0.0680	2.8913
2581-34-2	0.0	-25.889	0.0680	2.4465
5428-54-6	0.0	-24.768	0.0680	2.9709
51-28-5	0.0	-8.779	0.0680	6.6198
97-52-9	0.0	-14.933	0.0680	3.6874
96-96-8	0.0	-10.940	0.0680	4.2078
95-57-8	0.0	-26.627	0.0833	2.6693
106-48-9	2.0	-29.302	0.0833	3.1398
108-43-0	0.0	-29.208	0.0833	3.0165
95-56-7	0.0	-14.595	0.0833	3.2925
106-41-2	0.0	-17.545	0.0833	3.4832
591-20-8	0.0	-16.933	0.0833	3.4334
59-50-7	0.0	-36.210	0.0680	3.8056
615-74-7	0.0	-34.308	0.0680	3.3721
1570-64-5	0.0	-36.525	0.0680	3.7843
87-65-0	0.0	-32.186	0.0680	2.3766
95-77-2	0.0	-34.640	0.0680	2.9063
120-83-2	0.0	-33.137	0.0680	2.6095
576-24-9	0.0	-33.931	0.0680	2.0873
583-78-8	5.0	-33.242	0.0680	2.3302
591-35-5	0.0	-35.381	0.0680	2.8982
95-95-4	0.0	-38.237	0.0556	2.4830
609-89-2	0.0	-27.482	0.0556	4.3718
30171-80-3	0.0	-10.958	0.0556	9.4844
58-90-2	7.0	-41.677	0.0454	1.8011
1836-77-7	0.0	12.559	0.0833	8.6537
120-52-5	0.0	50.504	0.1021	19.7024
103-50-4	0.0	4.290	0.1021	11.1262
80-43-3	0.0	25.153	0.1021	17.1826
96-69-5	1.9	-75.994	0.0556	18.5044
117-81-7	43.0	-215.809	0.0833	12.0389
3648-21-3	36.0	-182.559	0.0833	10.2913
26761-40-0	42.0	-246.382	0.0833	13.7313
134-62-3	0.0	-13.900	0.0833	6.7786
98-73-7	0.0	-86.580	0.0833	6.1189
118-91-2	5.6	-66.762	0.0833	3.7789
121-92-6	0.0	-56.141	0.0833	4.4563
603-11-2	0.0	-136.242	0.0680	10.6115
91-15-6	3.3	87.699	0.0833	4.4901
623-26-7	0.0	86.143	0.0833	4.9026
626-17-5	0.0	86.162	0.0833	4.7756
1897-45-6	0.0	72.086	0.0370	1.4318
88-19-7	0.0	-50.818	0.0833	7.2016
88-44-8	0.0	-109.907	0.0680	7.2467
127-68-4#	0.0	-90.405	0.0833	9.4963
88-53-9	0.0	-115.916	0.0556	6.9987
946-30-5#	0.0	-92.167	0.0680	10.3243
330-55-2	0.0	-11.030	0.0680	7.5466
330-54-1	0.0	-9.652	0.0680	6.5388
24019-05-4	0.0	-21.302	0.0680	20.1781
2655-14-3	1.0	-60.739	0.0680	7.8530

3766-81-2	1.0	-65.010	0.0833	8.8373
2631-40-5	0.0	-58.812	0.0833	8.1476
644-97-3*	0.0	-28.527	0.1021	5.1768
1241-94-7*	2.0	-216.584	0.1021	24.0259
60628-17-3*	14.0	-188.863	0.1021	22.5740
26967-76-0*	0.0	-193.571	0.0833	37.1672
2597-03-7*	0.0	-144.500	0.1021	13.7624
122-14-5*	0.0	-132.895	0.0680	11.4238
2104-64-5*	3.0	67.154	0.0833	17.0437
80-54-6	8.0	-44.034	0.0833	7.8748
3319-31-1	4.2	-335.422	0.0680	14.7004
3886-69-9	6.0	15.621	0.1021	4.3026
782-74-1	0.0	75.952	0.0833	12.4101
1163-19-5	0.0	99.015	0.0370	18.8227
87-86-5	1.0	-44.715	0.0370	1.1598
36065-30-2	2.9	8.700	0.0556	8.7279
50-31-7	0.0	-75.048	0.0556	2.9728
10541-83-0	14.0	-67.225	0.0833	5.6621
98-10-2	0.0	-46.954	0.1021	6.3724
961-11-5*	0.0	-220.597	0.0556	10.7600
14816-18-3*	3.0	-64.576	0.1021	14.6704
1460-02-2	0.0	-32.100	0.0680	11.7982

*indicates phosphorus compound or highly influential compound which was removed
sodium salt; treated as undissociated acid
+o- and p-tricresyl phosphates respectively, although same CAS number given

3. Conclusions

Multiple linear regression analysis of 79 aromatic compounds found to be biodegradable in the MITI test did not produce satisfactory correlations. However, discriminant analysis based on three calculated parameters (one from MOPAC 7 and two from MOLCONN-X) gave good discrimination between readily and non-readily biodegradable compounds, with an overall correct cross-validated prediction rate of 78.6 %.

4. References

1. Dearden, J.C. and Nicholson, R.M. (1987). Correlation of biodegradability with atomic charge difference and superdelocalisability. In K.L.E. Kaiser (ed.), *QSAR in Environmental Toxicology - II*, D. Reidel, Dordrecht, pp.83-89.
2. Dearden, J.C. and Nicholson, R.M., unpublished work.
3. Boethling, R.S. (1986). Application of molecular topology to quantitative structure-biodegradability relationships. *Environ. Toxicol. Chem.* **5**, 797-806.
4. Boethling, R.S. and Sabljic, A. (1989). Screening level model for aerobic biodegradability based on a survey of expert knowledge. *Environ. Sci. Tech.* **23**, 672-679.
5. Dearden, J.C. (1996). Descriptors and techniques for quantitative structure-biodegradability studies. *SAR QSAR Environ. Res.* **5**, 17-26.

6. Dearden, J.C. and Nicholson,R.M. (1986). The prediction of biodegradability by the use of quantitative-structure-activity relationships: correlation of biological oxygen demand with atomic charge difference. *Pestic. Sci.* **17**, 305-310.

7. Dearden, J.C. and Stott, I.P. (1995). CoMFA analysis of biodegradability. *SAR QSAR Environ. Res.* **4**, 189-196.

8. Fujita,T., Nishioka, T., and Nakojima, M. (1977). Hydrogen bonding parameter and its significance in quantitative structure-activity studies. *J.Med. Chem.* **20**, 1071-1081.

9. Yang, G., Lien, E.J., and Guo, Z. (1986). Physical factors contributing to hydrophobic constant p. *Quant. Struct.-Act. Relat.* **5**, 12-18.

10. Hall, L.H., Mohney, B.K., and Kier, L.B. (1991). The electrotopological state: an atom index for QSAR. *Quant. Struct.-Act. Relat.* **10**, 43-51.

PREDICTION OF BIODEGRADABILITY FROM CHEMICAL STRUCTURE

Use of MITI I data, Structural Fragments and Multivariate Analysis for the Estimation of Ready and Not Ready Biodegradability

H. LOONEN, F. LINDGREN, B. HANSEN, W. KARCHER
European Commission, Joint Research Centre, Environment Institute
European Chemicals Bureau,
TP 280, 21020 Ispra (Va), ITALY

1. Abstract

Biodegradation is an important process in the environmental fate of substances. Therefore, biodegradation is considered for several legislative purposes, such as Priority Setting, Risk Assessment and Classification and Labelling of substances within the European Union. For these purposes, it is important whether a substance is readily or not-readily biodegradable according to EU and OECD test guidelines. For many existing substances, the experimental biodegradation data have been derived under non-standard test conditions which complicates the interpretation of test results. Furthermore, results from different studies may not be consistent. (Quantitative) Structure Activity Relationships ((Q)SARs) may assist in the evaluation of experimental data. Therefore, a model has been developed that predicts whether a substance is readily biodegradable or not-readily biodegradable under MITI I test conditions. The model is based on a set of 600 substances, all having MITI I test results and 111 predefined structural fragments. The model is generated by Partial Least Squares (PLS) discriminant analysis. The model is both internally cross validated and externally validated with 198 substances that were not used in the model development. Both readily and not-readily biodegradable substances were predicted correctly for >86 % of the substances in the external validation set. The model has been developed in line with the principles on Use of QSARs laid down in the EU Technical Guidance Documents on Risk Assessment.

2..Introduction

The environmental fate assessment of substances addresses the distribution of a substance in the environment, and its changes with time in concentration. Environmental fate assessment includes both biotic and abiotic transformation, where for many substances biotic degradation is the major way of removal from the environment. Therefore biodegradability is a key property for several legislative processes. In the European legislation the evaluation of "existing" substances with respect to their hazard and risk to the environment is covered by three steps: Priority Setting, Risk Assessment and Classification & Labelling of substances. All these steps take into account the

105

W. J. G. M. Peijnenburg and J. Damborský (eds.), Biodegradability Prediction, 105–113.
© 1996 *Kluwer Academic Publishers.*

biodegradability of a substance. The role of biodegradability in these processes is reviewed in the next section.

2.1.   BIODEGRADABILITY IN EU LEGISLATION

Priority Setting and Risk Assessment are covered by Council Regulation (EEC) 793/93, which provides the legal framework for the systematic evaluation of the so-called "existing" substances. Priority Setting of substances is required to indicate which substances need immediate attention because of their potential hazardous effects to man or the environment. The basis for Priority Setting is formed by an automated ranking method [1]. The automated ranking requires a.o. information on whether a substance is readily, inherently or not-readily biodegradable according to one of the standardised EU or OECD test guidelines for biodegradability [2,3]. For the aim of Priority Setting, conservative default values are assigned to substances for which no experimental data are available for the specified endpoints. However, for those endpoints where validated Structure-Activity Relationships (SARs) exist, the data gaps will be filled using these estimation techniques [1], thereby replacing the default values.

A comprehensive Risk Assessment is required for priority substances. The Environmental Risk Assessment of substances requires information on the rate with which substances are biodegraded in different environmental compartments. Compartments under consideration are waste water treatment plants, water, soil and sediment. Since measured data on degradation rate constants in the different compartments are usually not available, a procedure has been proposed to extrapolate these values from standardised ready and inherent biodegradation tests [4]. Experimental data on Ready Biodegradability should be supplied for all priority chemicals for which a risk assessment is carried out. Nevertheless, practice shows that the available data are often not in agreement with each other, and that information on Structure Biodegradation Relationships (SBRs) may give additional information for experimental data evaluation.

The Council Directive 67/548/EEC was originally adopted in order to provide uniform EU wide rules for the packaging, classification and labelling of dangerous chemicals. The 6th amendment to this Directive introduced the concept of "dangerous for the environment", and detailed criteria for the aquatic environment were adopted in the 12th adaptation to technical progress. Biodegradability is one of the parameters considered for Classification of substances as dangerous to the environment. Classification is performed on the basis of evaluated experimental data indicating whether a substance is readily or not-readily biodegradable. During the evaluation of experimental data, problems may arise concerning the availability, quality and consistency of biodegradation data. Again SBRs may assist in the evaluation of the experimental data.

It can be concluded that both Priority Setting, Environmental Risk Assessment and Classification of substances address biodegradability in terms of ready, inherent or not-ready biodegradation according to the principles laid down in the standardised EU and OECD test protocols for determination of biodegradability. The three processes are often hampered by the lack of consistent experimental biodegradation data. In these circumstances SBRs may serve as a supporting tool in data evaluation. In order to be suitable for these purposes, the SBRs need to be compatible with the endpoints for biodegradability used in the legislation and they need to be applicable to a wide variety of structures.

2.2.   VALIDATION STUDIES OF EXISTING SBRs

Several reviews on SBRs have been performed recently [5,6,7]. OECD [5] reviewed and evaluated 78 existing models, and derived some new ones. The SBRs were validated by comparison of the predictions made by the respective models with experimental data determined according to the MITI I test procedure [2] for more than 700 substances. It was concluded that only few existing SBRs provided an adequate level of agreement between estimated and experimental data. The limited applicability could be traced back to lack of endpoint homogeneity in the training set, inconsistent test data, and use of restricted data sets [5].

In the EU supported project "QSAR for predicting fate and effects of chemicals in the environment", 84 models for biodegradation were evaluated, including 31 qualitative and 53 quantitative models [6,7]. Different kind of descriptor variables were used in the models. Structural fragment contribution methods turned out to be the most successful way of modelling biodegradability. However, the authors stated that strictly spoken, none of the models meet the selection criteria applied in this project [6]. It appeared that the majority of models could not be recommended for use for legislative purposes, because many models were based either on not compatible biodegradation endpoints, or too restricted data-sets. The "best available" models were considered to be two models described in the OECD report [5] for respectively acyclic compounds and mono substituted benzenes and the general applicable BIODEG program [8]. These models were validated with a set of 488 MITI I test data. It was concluded that the models for acyclic compounds and mono substituted benzenes performed rather well, with all predictions being correct for over 75 % of the compounds tested. The overall percentage of correct predictions (slow and fast biodegradation) made by the BIODEG programs was 56 % and 63 % for respectively the BIODEG Probability Programs 1 and 2 (BPP1 and BPP2). The percentage of correct "fast biodegradation" predictions was low (< 40 %), whereas the "slow biodegradation" predictions were very accurate (> 90 % correct) [7]. Therefore, it was recommended to use this program only in a conservative way (i.e. only the slowly degradable predictions).

2.3.   OBJECTIVE OF NOVEL MODEL DEVELOPMENT

The objective of the study described in the current paper, was the development of a model that can predict accurately both readily and not-readily biodegradability. The predictions should be coherent with the endpoint of biodegradability according to standardised EU and OECD biodegradation test guidelines. Consequently, the model predictions would be compatible with the endpoint of biodegradation used for Priority Setting, Environmental Risk Assessment and Classification & Labelling. The model is based on MITI I test data for a wide variety of substances, making the model applicable to a large range of chemical structures. A set of 111 structural fragments is used as model descriptors. The structural fragments are not limited to the known rules of thumb, regarding structure-biodegradation relationships. In contrast, the aim of using this set of structural fragments was a general chemical characterisation of the substances, which gives the opportunity to explore areas of SBR, which have not been investigated so-far. The model is calculated by Partial Least Squares (PLS) discriminant analysis. The model was externally validated with a set of 198 substances, which were not used for the model generation. In order to assure that the model can be used within the EU legislative framework, the model has been developed and validated in agreement with the principles on use of QSARs, laid down in the EU Technical Guidance Documents on Risk Assessment [4].

3. Methods

3.1.   EXPERIMENTAL DATA

The model is based on experimental biodegradation data determined according to the MITI I test protocol. The MITI I test was developed in Japan, and at present it constitutes one of the six standardised ready biodegradability tests described by EU and OECD (EU test code C.4-F and OECD test code 301 C [2,3]). The MITI I test is a method that permits the screening of chemicals for ready biodegradability in an aerobic aqueous medium. For the MITI I test, 100 mg/l of test substances is inoculated and incubated with 30 mg/l sludge. BOD is measured continuously during the test period. The pass level for ready biodegradability is reached, if the BOD amounts $\geq$ 60 % of ThOD. This pass level has to be reached within the 28 day test period, the test can be stopped after a shorter test duration, if the pass level is reached after a shorter test period (e.g. 2 or 3 weeks).

For the model development, experimental data determined according to the MITI I test protocol were collected for 798 substances. All experimental data were determined by the Chemicals Inspection & Testing Institute (CITI) of Japan. Part of the data had been published [9], the others were received directly from CITI.

3.2.   STRUCTURAL FRAGMENTS

The descriptor variables used in the model development consist of 149 structural fragments. This set of structural fragments had originally been developed to allow the retrieval of chemicals from large data bases, on the basis of substructure searches [10].

The chemical characterisation by structural fragments is readily generated by computer programs, originally developed for ICI and later adapted for the Joint Research Centre of the European Commission [10].

3.3.   MULTIVARIATE ANALYSIS

The model was calculated by Partial Least Squares (PLS) discriminant analysis. PLS is a projection method that relates the information in the biodegradation data (Y-variable) to the systematic variation in the structural fragments matrix (X-variables), using latent vectors [11,12,13]. Standard calculation procedures were applied: the variables were mean centred and scaled to unit variance. The model was calculated using the SIMCA software [14]. The model was generated on a set of 600 substances, which were randomly selected from the MITI I data base. The remaining 198 substances were used for external validation of the models.

3.4. MODEL VALIDATION

Much attention has been paid to the validation of the predictive capability of the model by both internal and external validation. Validation has been performed in agreement with the principles on evaluation of the accuracy of QSARs, as described in the EU Technical Guidance Documents on Risk Assessment [4]. Internal validation was performed by cross-validation, resulting in the cross-validated correlation coefficient Q^2. External validation was performed by comparison of the predictions made by the model for 198 substances with known MITI I data. The results of the external validation are expressed in terms of percentages of correct predictions.

4. Results and Discussion

4.1. STRUCTURE-BIODEGRADABILITY MODEL

A structure-biodegradability relationship is calculated by PLS discriminant analysis. The structural fragments are used as X-descriptors and the biodegradation data as the dependent Y-variable. The total set of structural fragments consisted of 149 different substructures, of which 111 actually occurred in the set of 600 substances used for the SBR calculation. The first PLS component described 41 % of the variation in the Y-variable, the biodegradation data. The corresponding cross-validated result was 35 %. With additional PLS components, the explained variation in the Y-variable increased, but the cross-validated regression coefficient remained the same, indicating that the predictive capability of the model did not increase with additional components.

At first glance, an explained variance in the Y-variable of 41 % might seem low. It should however be considered that the Y-variable is of a binary nature: the substance is either readily (indicated by 1) or not-readily biodegradable (indicated by 0) and these values are subsequently compared to a continuous scale (ranging from -0.62 to 1.38). More important to notice is that the cross validated regression coefficient Q2 is 85 % of the regression coefficient R^2, indicating that most of the explained variance can also be predicted.

Figure 1 shows a plot of the t1 - t2 scores, displaying the observations as situated on the projection plane. The plot reveals a separation between the readily and not-readily biodegradable substances. The substances that are readily biodegradable in the experimental tests are situated in the upper right area of the plot, while the substances that did not pass the ready criteria in the experimental test, are located in the lower left area of the plot. In the centre, there is a border of overlap between the readily and not-readily biodegradable substances. Some substances were predicted in the wrong area. It concerned mainly substances that were readily biodegradable in the experimental test, but which were clearly located in the area of not-readily biodegradable substances by the model. The wrong predictions were mostly aromatic compounds with one or more

110

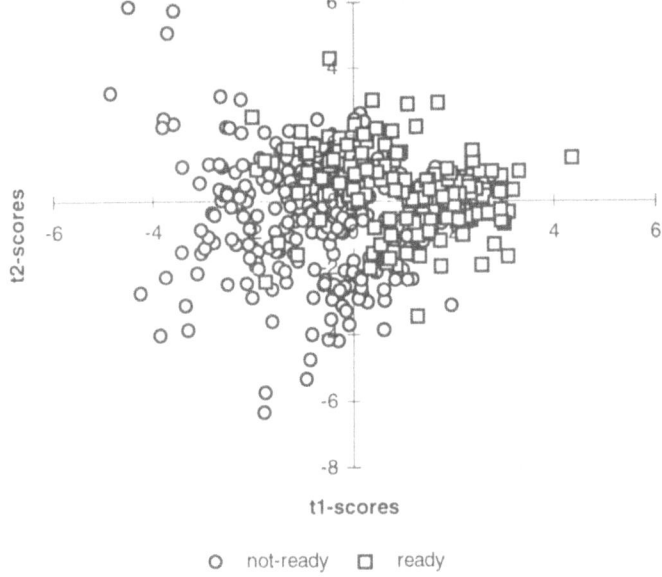

Figure 1.   Plot of t2 versus t1 scores calculated by PLS for 600 substances in the training set.

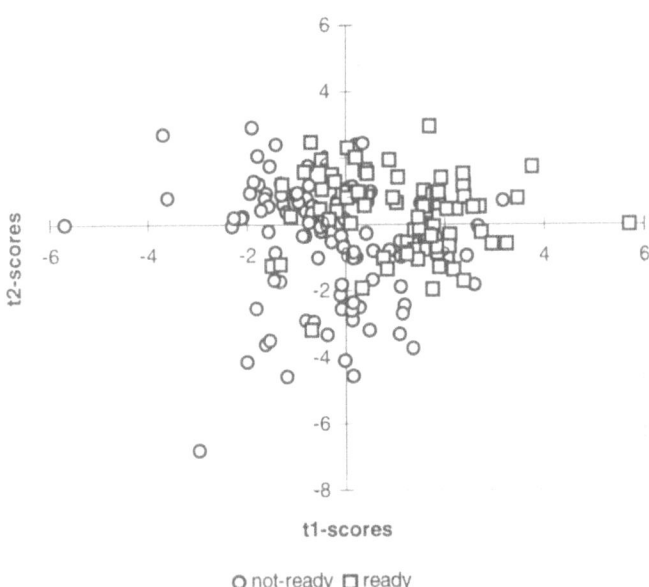

Figure 2.   Plot of t2 versus t1 scores calculated by PLS for 198 substances in the external validation set.

halogenated alkane substituents, predicted to be not-readily biodegradable. Examples of these wrong predictions are trichloromethyl benzene, dichloromethyl benzene, and monochloromethyl benzene.

4.2. EXTERNAL VALIDATION OF MODEL

The model was externally validated with a set of 198 MITI I data which were not used for the model development (testing set). For the aim of external validation, predictions for the 198 substances were calculated by the model. Figure 2 displays the results of these calculations. Again a separation is visible, with the readily biodegradable substances in the upper right part and the not-readily biodegradable substances in the lower left part of the plot.

In order to make a comparison between the binary experimental data and the predictions on a continuous scale, a transformation of the predictions was necessary. A predicted value less than 0.55 was considered as not-readily biodegradable; a predicted value greater than 0.55 was considered as readily biodegradable. With this transformation all predictions were taken into account inclusive predictions in the borderline area. An alternative transformation was applied by excluding 10 % of the substances, which were in the borderline area between ready and not-ready biodegradation (0.50 - 0.65).

The overall results of the validation studies are summarised in table 1. The results are summarised in terms of the percentage of predictions that is correct, which is considered to be the most important parameter from a users point of view.

From table 1, it can be seen that the overall predictions made by the PLS model are correct for 83 % of the substances in the external validation set (including borderline substances). If 10 % of the borderline predictions are excluded the percentage of correct predictions increases to 87 %. Ready and not-ready predictions are equally correct (86 % and 88 % respectively).

TABLE 1.   Results of the external validation study: comparison between the newly developed PLS model and the BIODEG Probability Program. The percentage of correct predictions is indicated. The external validation set consisted of 198 substances.

	PLS model % correct predictions of all predictions	PLS model % correct predictions excl. borderline	BIODEG % correct predictions excl. borderline
Not Ready	84 %	88 %	90 %
Ready	82 %	86 %	59 %
overall R + NR	83 %	87 %	67 %

3.3. COMPARISON WITH BIODEG PROBABILITY PROGRAM

The BIODEG Probability Program (BPP) [8] has been indicated to be the "best available" model for prediction of biodegradability in a recent review study [6,7]. The same kind of validation was performed for this program, using the same set of 198 substances. This program constitutes of 4 models (BPP1-4), of which the relevant BPP1 and BPP2 were used. If BPP1 and BPP2 predicted controversial results (8 % of the predictions), the predictions were not taken into account. The predictions were interpreted as indicated by the model developers [8]. The results are also included in table 1.

For the BIODEG PROBABILITY PROGRAM, the overall percentage of correct predictions is 67 %. A clear difference is found between the correct percentages of fast and slow predictions. The predictions of slow biodegradation are highly reliable, with 90 % correct predictions. In contrast, the prediction of fast biodegradation is often not in agreement with the experimental data (59 % correct). These percentages are exclusive 8 % of the substances which are considered to be borderline cases.

5. Conclusions and Outlook

The model presented is able to predict readily and not-readily biodegradability correctly for 84 % of the substances. It can be concluded that this model is more accurate, than other comparable models published so far in literature. The model is fully compatible with the endpoint for biodegradability used for legislative purposes. Currently, investigations are made to further define and optimise the predictive capability of the model. To this end, the training set of the model will be extended with MITI I data for additional substances. Apart from individual structural fragments, the influence of fragment-fragment interactions will also be included in the model. The PLS modelling will be optimised. The model will be subjected to further thorough validation. The structural domain and the meaning of the structural fragments and their interactions will be investigated in detail.

6. Acknowledgements

Dr. M. Takatsuki (CITI, Japan) and Dr. W. Peijnenburg (RIVM, The Netherlands) are gratefully acknowledged for kindly providing MITI I data to the European Commission. Furthermore, we thank Dr. J. Niemelä (NERI, Denmark) for providing the BIODEG Probability Program estimates.

7. References

1. European Commission (1996) Priority Setting using the IPS method in support of Council Regulation (EEC) 793/93, European Chemicals Bureau doc. 4/05/96 (in prep.).
2. European Commission (1992) Methods for the determination of ecotoxicity, *Official Journal of the European Communities* No L 383 A/ 187-225.

3. OECD (1994) OECD Guidelines for the testing of chemicals. Guideline no. 301 Ready Bio-degradability, Organisation for Economic Cooperation and Development, Paris.
4. European Commission (1996) EU Technical Guidance Documents for risk assessment of new and existing substances, European Commission, Luxembourg.
5. Degner, P., Müller, M., Nendza, M., and Klein, W. (1993) Structure-activity relationships for biodegradation. OECD Environment monographs No. 68, Paris.
6. Rorije, E., Langenberg, J.H., and Peijnenburg, W.J.G.M. (1995). QSARs for biodegradation. In: Overview of structure-activity relationships for environmental endpoints, Part 1: General outline and procedure. Hermens, J.L.M. (Ed.), Report prepared within the framework of the project "QSAR for Prediction of Fate and Effects of Chemicals in the Environment", an international project of the Environmental Technologies RTD Programme (DGXII/D-1) of the European Commission under contract number EV5V-CT92-0211.
7. Langenberg, J.H., Peijnenburg, W.J.G.M., and Rorije, E. (1995) On the usefulness and reliability of existing QSBRs for risk assessment and priority setting, *Sar QSAR Environ. Res.* in press.
8. Howard, P., and Meylan, W. (1992) Biodegradation Probability Program, Version 3. Syracuse Research Corporation, NY.
9. Chemicals Inspection & Testing Institute Japan (1992) Data of existing chemicals based on the CSCL Japan, Japan Chemical Industry Ecology-Toxicology & Information Center, Japan.
10. Eakin, D.R., Hyde, E., and Palmer, G. (1974) The use of computers with chemical structural information: ICI CROSSBOW System, *Pestic. Sci.* **5**, 319-326.
11. Wold, S., Martens, H. and Wold, H. (1983) The multivariate calibration problem in chemistry solved by the PLS method, in A. Ruhe and B. Kagstrom (eds), *Matrix Pencils*, Springer Verlag, Heidelberg, pp. 286-293.
12. Garthwaite, P.H. (1994) An interpretation of Partial Least Squares, *Journal of the American Statistical Association* **89**, 122-127.
13. Lindgren, F., Nouwen, J., Loonen, H., Worth, A., Hansen, B., Karcher, W. (in press) Environmental modelling based on a structural fragments approach, in Hoskins, J.A. (ed.), proceedings Assessing low-dose effects of air pollutants, *Indoor & Built Environment*, Karger medical and scientific publishers, Basel.
14. SIMCA-software, Address: Umetri AB, S-901 24 Umea, Sweden.

DEVELOPMENT OF STRUCTURE BIODEGRADABILITY RELATIONSHIPS (SBRs) FOR ESTIMATING HALF-LIVES OF ORGANIC CONTAMINANTS IN SOIL SYSTEMS

R. GOVIND, and L. LEI
Department of Chemical Engineering, University of Cincinnati
Cincinnati, OH 45221-0171, USA

and

H. TABAK
USEPA, National Risk Management Research Laboratory, ORD
26 W. Martin Luther King Drive, Cincinnati, OH 45268, USA

1. Abstract

Knowledge of half-lives or biodegradation rate constants in soil is useful for estimating the natural attenuation rates of contaminants due to microbial transformations and to make decisions regarding treatment action or no treatment with isolation of the contaminated site to minimize exposure to animal and human life. Half-life is defined as the time required for 50 % of the contaminant to be biodegraded. Soil treatment is time consuming and expensive, and often for large isolated contaminated sites, relying on natural attenuation may be the most cost-effective solution. In this paper, a neural network is trained to estimate the range of half-lives for organic contaminants in soil. Soil half-life data, obtained from the literature for 258 chemicals is correlated with 14 molecular fragments or indicators using a back-propagation neural network with 14 input nodes, 12 nodes in the hidden layer and 2 output nodes. A cross-validation method was used to test the neural network. The converged neural network produced less than 50 % relative error for more than 80 % of the chemicals in the training set. Using a classification scheme of fast (half-life range of 1 to 7 days), moderately fast (half-life range from 7 to 28 days), slow (half-life range from 28 to 180 days) and resistant (half-life range from 180 to 365 days), the neural network was able to correctly classify more than 95 % of the 258 chemicals in the database.

2. Introduction

Biodegradation is an important breakdown mechanism that occurs due to a variety of naturally present microorganisms in aquatic and soil ecosystems. Many factors, such as contaminant properties, temperature, pH, type of soil, microbial population density, etc., can influence the rate and extent of biodegradation. The ability to predict biodegradation rates can enhance our ability to design synthetic organic chemicals with

115

W. J. G. M. Peijnenburg and J. Damborský (eds.), Biodegradability Prediction, 115–138.
© 1996 *Kluwer Academic Publishers.*

desired properties and favorable environmental impact. Development of structure-bio-degradability relationships (SBRs) is valuable in minimizing or preventing contaminant persistence in the environment.

Biodegradation in soil is a complex process and although recent studies attempt to quantitate biodegradation kinetics in various types of soil reactors, there is no extensive database on biodegradation kinetic rates in soil for a majority of typical organic contaminants. Further, experimental measurement of biodegradation rates in soil is time consuming, labor intensive and expensive. Knowledge of soil biokinetics for various contaminants is essential for determining treatment times and effectiveness of biological processes for treating contaminated soil sites as well as determining the effectiveness of using natural bio-attenuation as a treatment strategy.

Many researchers, summarized by Howard et al. [1], have attempted to correlate physical properties and chemical structure with biodegradation activity. Models that have been developed include discriminant function analysis [2], molecular connectivity indexes [3], group contribution approaches [4,5] and linear and non-linear regression equations [1].

Quantitative structure-activity relationships (QSARs) have been developed using neural networks for estimating aqueous biodegradation kinetics [6], compound toxicity [7], and recently soil partitioning coefficient [8]. In some of these approaches, the molecular structure is decomposed into fragments or groups and a unique contribution towards the final parameter value is assigned to each fragment (group). The group contribution methods can be used for a wide variety of chemicals, since a large number of chemicals can be constructed from a few groups and this approach also allows systematic design of new molecular structures with a desired value of the final parameter.

Structure-biodegradability relationships (SBRs) are predominantly based on chemicals that are biodegradable and hence have limited capability to predict chemical recalcitrance. Further, chemical transformations are easier to predict than microbial conversions [9]. One major problem is the variety of natural microbiota that are capable of degrading a chemical and a general lack of standard protocols for quantitating biodegradation rates in aqueous and soil ecosystems. Recently, attempts have been made to develop experimental techniques and mathematical models for measuring intrinsic biodegradation rate parameters for aqueous and soil systems [10,11].

In this paper, biodegradation rates in soil, expressed as half-lives, are correlated with molecular groups using a neural network. There is a simple relationship between the half-life ($t_{1/2}$) and the first-order biodegradation rate constant (k):

$$t_{1/2} = 0.693 / k \qquad (1)$$

The first-order biodegradation rate constant (k) can be related to the contaminant concentration, as follows:

$$-\frac{d[C]}{dt} = k[C] \tag{2}$$

where [C] is the contaminant concentration, t is time, and k is the first-order biodegradation rate constant. Since half-lives in soil are expressed as a range, both lower and upper bounds are used in our analysis. Soil half-life data, obtained from the literature [12], for 258 chemicals is correlated with 13 molecular fragments or indicators. One indicator (Indicator #14), which is a binary variable, is assigned a value of 1 if the molecular structure is symmetric, otherwise it has a value of zero. The 258 compounds, used in this study, are structurally diverse and consist of aliphatic hydrocarbons, cyclic hydrocarbons, halogenated hydrocarbons, benzene and benzenoid compounds, polycyclic aromatic hydrocarbons (PAHs), nitrofurans and others. Molecular fragments or groups are selected only if they appear in at least 20 compounds, i.e., in at least 10 % of the total number of chemicals in the training set. Further, selection of molecular groups was also based on literature studies wherein certain groups have been associated with biodegradation rates (presence of hydroxyl, aldehyde, carboxyl, ester, and amide functional groups result in higher biodegradation rates; hydroxyl and carboxyl groups on benzene rings usually increase biodegradation rates; increase in the number of chlorine atoms decreases biodegradation rates, etc.).

There is an extensive literature on mathematical models of artificial neural networks [13,14,15]. Neural networks have found applications in image and speech recognition, on-line diagnosis of process faults, process control and optimization of complex functions. Artificial neural networks consist of many non-linear computational elements operating in parallel and arranged in patterns similar to biological neural nets. The nodes or computational elements are connected via weights that are typically adapted during use to improve performance. Superior performance is achieved by using dense interconnection of simple computational elements.

The neural network is trained using a back-propagation algorithm, which uses a gradient search technique to minimize a cost function equal to the mean square difference between the desired and actual net outputs. The network is trained by initially selecting small random weights and internal thresholds and then presenting the training data for each chemical. Weights are adjusted after each trial, until the weights converge and the cost function is reduced to an acceptable value.

3. Structure of Input Data

Each chemical used in this analysis was represented by a set of molecular indicators. A molecular fragment is not chosen as an indicator until it appears in at least 20 chemicals in the training database. Further, selection of indicators was also based on what molecular fragments have been used in structure-biodegradability relationships (SBRs)

TABLE 1.   List of molecular indicators used for describing each molecular structure and the number of chemicals from the database in which each indicator appears.

Name	Fragment included	Value of fragment	Class of chemicals	Number of chemicals involved in database
Indicator #1	-CH3 -(CH2)- -(CHR)- -(CR2)-	1 1 2 3	Alkyl	169
Indicator #2	-CHO	1	Aldehyde	47
Indicator #3	-C(O)- (structure)	1 1	Carboxyl	40
Indicator #4	-COO- -COOH	1 1	Ester, carboxylic acid	50
Indicator #5	-OH (structure)	1 1	Alcohol	41
Indicator #6	-HC=CH- -(C=)-	1 1	Double bounds	93
Indicator #7	- Cl -F -Br -I	1 1 1 1	Halogen	90
Indicator #8	-NH- -NH2- =N- -NO2	1 1 1 1	Nitrogen containing	103
Indicator #9	(benzene ring)	1	Ar : substituted or unsubstituted aromatic	146

TABLE 1.   List of molecular indicators used for describing each molecular structure and the number of chemicals from the database in which each indicator appears (continued)

Name	Fragment included	Value of fragment	Class of chemicals	Number of chemicals involved in database
Indicator #10			substituted or unsubstituted two ring PAHs	23
		1		
Indicator #11			substituted or unsubstituted three or more ring PAHs	24
		1		
		2		
Indicator #12		1	substituted cyclic hydrocarbons	35
Indicator #13		1	cyclic hydrocarbons	45
Indicator #14		1/0	Symmetric Structure	102

presented in the literature.   For example, it is known that presence of hydroxyl, aldehyde, carboxyl, ester and amide molecular groups enhances biodegradation rates whereas presence of chlorine atoms decreases biodegradation rates.   Hence, these molecular fragments were used as indicators in this study.

Table 1 lists the molecular indicators used in describing each molecular structure and the number of chemicals from the database in which each indicator appears.   The value assigned to each indicator, except for Indicator #14, are additive for a molecular structure.   Indicator #14 is an indication of a symmetric molecular structure, and is an

Figure 1. Two examples of input chemical structure representation using an input string based on the 14 indicators shown in Table 1.

Example

(1) The strúcture is

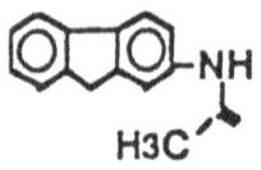

(2)

fragment	quantity	value of fragment	indicator #	values of indicator
	2	1	#9	2
	1	1	#13	1
-NH-	1	1	#8	1
-(C=O)-	1	1	#6	1
-CH3	1	1	#1	1

(3) Input:

1 0 0 0 0 1 0 1 2 0 0 0 1 0

Example   DDT

(1) The structure is

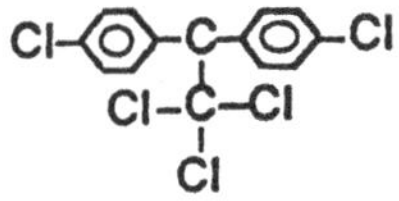

(2)

fragment	quantity	value of fragment	indicator #	values of indicator
-Cl	5	1	#7	5
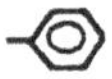	2	1	#9	2
-(CHR)-	1	1	#1	4
symmetry			#14	1

(3) The input:

1 0 0 0 0 0 5 0 2 0 0 0 0 1

important input variable, since symmetric molecular structures consistently exhibited longer half-lives in soil, i.e., were slower to biodegrade.

There are three basic steps involved in entering a chemical in the database or using the approach to obtain an estimate of half-life in soil: (1) Examine the chemical's molecular structure; (2) Identify the molecular fragments (indicators) from Table 1 which are present in the chemical structure and calculate the value for each indicator; If a molecular fragment is not listed in Table 1, use a structurally similar indicator from Table 1 to obtain an estimate using the approach; and (3) Construct a string of input numbers $\{n_I, I = 1,2,3 ...14\}$, where n_I is the value for each indicator.

For example, Allyl alcohol ($H_2C=CHCH_2OH$) has the following molecular fragments:
(1) $\{-(C=)-\}$; (2) $\{-CH_2CH_2-\}$; and (3) $\{-OH\}$. Each molecular fragment appears only once in the chemical structure and has a value of 1. The three molecular fragments, are represented as the following indicators, in Table 1:   Indicator # 1: $\{-CH_2CH_2-\}$; Indicator # 5: $\{-OH\}$; and Indicator # 6: $\{-(C=)-\}$. Based on this representation, allyl alcohol is represented by the following input string: $\{1\ 0\ 0\ 0\ 1\ 1\ 0\ 0\ 0\ 0\ 0\ 0\ 0\ 0\}$ where 1 represents the presence and value of the indicator, represented by the position of the number in the input string and 0 represents the absence of the indicator in the molecular structure.   Figure 1 shows two other examples of molecular structures and the input strings that are used to represent them in our analysis.

From the 258 chemicals entered in the database, shown in Table 2, each represented by a string of numbers $\{n_I\}$, 248 chemicals were used for training the neural network, and 10 chemicals were used for testing the neural network. Data reported for half-life for each chemical in the training set, the natural logarithm of both the upper and lower values, were used in the training of the neural network. Data for chemicals in the testing set were not used for training the neural network and were selected using the following reasoning: (1) these chemicals were different in structure and all 14 indicators were involved; and (2) they varied considerably in the values of the half-lives.

4. Description of the Neural Network

A back-propagation neural network with 14 input nodes, 12 nodes in the hidden layer and 2 output nodes, is used to correlate the input data with the log values of the high and low soil half-lives for each chemical in the database.   The back-propagation network maps the given input data with the desired output values by minimizing a system error, defined as a total squared error. The network consists of three layers: input, hidden and output layer. In the network, input information is passed forward until a set of output values are calculated. The error between the calculated and desired output values are then passed backwards to modify the network connection weights starting with the weights between the output and hidden layers.

The neural network was tested using the "cross-validation" method. Instead of dividing the total number of chemicals in the database into training and testing groups, ten chemicals are randomly picked as test compounds and the neural network is trained using the rest of the chemicals in the database. Training is ended when the total squared error is below a minimal value. The test compounds are then entered in the neural

TABLE 2. Listing of 258 chemicals in the database, with the low and high values of the half-lives in days, and the input strings used to represent each chemical in training the neural network.

1 Acenaphthene	102 12.3	0 0 0 0 0 0 0 0 2 3 0 1 0 0
2 Acenaphylene	60 42.5	0 0 0 0 0 1 0 0 3 1 0 0 1 0
3 Acetamide	7 1	1 1 0 0 0 0 0 1 0 0 0 0 0 0
4 Acetone	7 1	2 1 0 0 0 0 0 0 0 0 0 0 0 0
5 Acetonitrile	28 7	1 0 0 0 0 1 0.0 0 0 0 0 0 0
6 2-Acetylaminofluorene	180 28	1 1 0 0 0 0 0 1 2 0 0 0 1 1
7 Acrolein	28 7	0 1 0 0 0 1 0 0 0 0 0 0 0 0
8 Acrylic-acid	7 1	1 1 0 0 0 1 0 1 0 0 0 0 0 0
9 Acrylonitrile	23 1.3	0 0 0 0 0 3 0 0 0 0 0 0 0 0
10 AflatoxinB1	28 7	0 2 0 1 0 4 0 0 1 0 0 3 4 0
11 Aldicarb	361 20	4 0 1 1 0 1 0 2 0 0 0 0 0 1
12 Alfrin	592 21	0 0 0 0 0 2 6 0 0 0 0 0 4 1
13 Allyl-alcohol	7 1	1 0 0 0 1 1 0 0 0 0 0 0 0 0
14 1-Amino-2-methylanthraquinone	28 7	1 2 0 0 0 0 1 2 0 0 0 1 0
15 2-Aminoanthraquinone	28 7	0 2 0 0 0 2 0 1 2 0 0 0 1 0
19 o-Anisidine	180 28	1 0 0 1 0 0 0 1 1 0 0 0 0 1
20 p-Anisidine	28 7	1 0 0 1 0 0 0 1 1 0 0 0 0 0
21 Anthracene	460 50	0 0 0 0 0 0 0 0 3 2 1 0 0 1
22 Auramine	180 28	4 0 0 0 0 1 0 3 2 0 0 0 0 1
23 Azaserine	56 8.2	1 0 2 1 0 1 0 2 0 0 0 0 0 0
24 Aziridine	28 7	0 0 0 0 0 0 0 1 0 0 0 1 1 0
25 Benefin	120 21	2 0 0 0 0 2 3 3 1 0 0 0 0 0
26 Benz(a)anthracene	680 102	0 0 0 0 0 0 0 0 4 3 2 0 0 1
27 Benz(c)acridine	365 180	0 0 0 0 0 0 0 0 4 3 2 1 0 1
28 Benzamide	15 2	0 1 0 0 0 0 0 1 1 0 0 0 0 0
29 Benzene	16 5	0 0 0 0 0 0 0 0 1 0 0 0 0 1
30 1,3-Benzenediamine	28 7	0 0 0 0 0 0 2 1 0 0 0 0 0
31 Benzidine	8 2	0 0 0 0 0 0 2 2 0 0 0 0 0 0
32 Benzo(a)pyrene	530 57	0 0 0 0 0 0 0 0 5 6 3 0 0 1
33 Benzo(b)fluoranthene	610 360	0 0 0 0 0 0 0 0 4 2 1 0 1 1
34 Benzo(ghi)perylene	650 590	0 0 0 0 0 0 0 0 5 4 3 0 1 1
35 Benzo(k)fluranthene	2140 910	0 0 0 0 0 0 0 4 2 0 0 1 1
36 Biphenyl	7 1.5	0 0 0 0 0 0 0 0 2 0 0 0 0 0
37 Bis(2-choloethyl)ether	180 28	2 0 0 1 0 0 2 0 0 0 0 0 0 1
38 Bromoethylene	180 28	0 0 0 0 0 1 1 0 0 0 0 0 0 1
39 Bromoform	180 28	1 0 0 0 0 0 3 0 0 0 0 0 0 1
40 1,3-Butadiene	28 7	2 0 0 0 0 2 0 0 0 0 0 0 0 0
41 Butanol	7 1	2 0 0 0 1 0 0 0 0 0 0 0 0 0
42 sec-Butyl-alcohol	7 1	2 0 0 0 1 0 0 0 0 0 0 0 0 0
43 tert-Butyl-alcohol	200 15	4 0 0 0 1 0 0 0 0 0 0 0 0 1
44 Butyl-benzyl-phthalate	7 1	5 0 1 3 0 0 0 0 2 0 0 0 0 0
45 1,2-ButyleneOxide	12.9 7	1 0 0 0 0 0 0 0 0 0 0 0 1 0
46 Butylglycolyl-butyl-phthalate	7 1	2 0 3 3 0 0 0 0 1 0 0 0 0 0
47 C.I.BasicGreen4	180 28	4 0 0 0 0 2 2 0 3 0 0 0 0 1
48 C.I.Solvent-Yellow3	28 7	2 0 0 0 1 0 3 2 0 0 0 0 0 0
49 C.I.Solvent-Yellow14	28 7	0 0 0 0 1 1 0 2 3 2 0 0 0 0
50 C.I.Vat-Yellow-4	180 28	0 2 0 0 0 3 0 0 4 0 0 0 2 1
51 Captan	60 2	0 2 0 0 0 1 3 1 0 0 0 1 1 0
52 Carbon-tetrachloride	265 180	1 0 0 0 0 0 4 0 0 0 0 0 0 1
53 Catechol	7 1	0 0 0 0 2 0 0 0 1 0 0 0 0 0
54 Chlordane	1386 238	0 0 0 0 0 1 8 0 1 0 0 0 4 1
55 Chloroacetic-acid	7 1	1 0 1 0 0 0 1 0 0 0 0 0 0 0
56 2-Chloroacetophenone	28 7	1 1 0 0 0 0 1 0 1 0 0 0 0 0

TABLE 2. Listing of 258 chemicals in the database, with the low and high values of the half-lives in days, and the input strings used to represent each chemical in training the neural network (continued).

57	Chlorobenzene	150 68	0 0 0 0 0 0 1 0 1 0 0 0 0 1
58	Chlorobenzilate	35 7	2 0 1 1 1 0 2 0 2 0 0 0 0 0
59	Chloroethane	28 7	1 0 0 0 0 0 1 0 0 0 0 0 0 0
60	Chloroform	180 28	1 0 0 0 0 0 3 0 0 0 0 0 0 1
61	Chloroprene	180 28	1 0 0 0 0 1 1 0 0 0 0 0 0 1
62	3-Chloropropene	14 6.9	1 0 0 0 0 1 1 0 0 0 0 0 0 0
63	Chrysene	1000 371	0 0 0 0 0 3 0 0 4 3 2 0 0 1
64	p-Cresidine	180 28	2 0 0 1 0 0 0 1 1 0 0 0 0 1
65	o-Cresol	7 1	1 0 0 0 1 0 0 0 1 0 0 0 0 0
66	Crotonaldehyde(trans)	7 1	1 1 0 0 0 1 0 0 0 0 0 0 0 0
67	Cumene	8 2	2 0 0 0 0 0 0 0 1 0 0 0 0 0
68	Cumene-hydroperoxide	28 7	2 0 0 1 1 0 0 0 1 0 0 0 0 0
69	Cupferron	180 28	0 0 0 0 0 1 0 2 1 0 0 0 0 1
70	Cyclophosphamide	28 7	2 1 0 0 0 0 2 2 0 0 0 1 0 0
71	2,4-DB	7 1	1 0 1 2 0 0 2 0 0 0 0 0 0 0
72	DDD	5833 730	1 0 0 0 0 0 4 0 2 0 0 0 0 1
73	DDT	5833 730	1 0 0 0 0 0 5 0 2 0 0 0 0 1
74	Dalapon	60 14	1 0 1 0 0 0 2 0 0 0 0 0 0 0
75	Di-n-butyl-phthalate	23 2	6 0 2 2 0 0 0 0 1 0 0 0 0 0
76	Diallate	90 10.5	5 1 0 0 0 1 2 2 0 0 0 0 0 0
77	2,4-Diaminotoluene	180 28	1 0 0 0 0 0 2 1 0 0 0 0 0 1
78	Diaminotoluenes	180 28	1 0 0 0 0 0 2 1 0 0 0 0 0 1
79	Dibenz(a,h)anthracene	940 361	0 0 0 0 0 0 0 5 4 3 0 0 1
80	Dibenzofuran	28 7	0 0 0 0 0 2 0 0 2 0 0 1 1 0
81	1,2,7,8-Dibenzopyrene	361 232	0 0 0 0 0 0 0 6 7 4 0 0 1
82	Dibromochloromethane	180 28	1 0 0 0 0 0 3 0 0 0 0 0 0 1
83	m-Dichlorobenzene	180 28	0 0 0 0 0 0 2 0 1 0 0 0 0 1
84	1,2-Dichlorobenzene	180 28	0 0 0 0 0 0 2 0 1 0 0 0 0 1
85	p-Dichlorobenzene	180 28	0 0 0 0 0 0 2 0 1 0 0 0 0 1
86	3,3'-Dichlorobenzidine	180 28	0 0 0 0 0 0 2 2 2 0 0 0 0 1
87	Dichlorodifluoromethane	180 28	1 0 0 0 0 0 4 0 0 0 0 0 0 1
88	1,1-Dichloroethane	154 32	2 0 0 0 0 0 2 0 0 0 0 0 0 0
89	1,2-Dichloroethane	180 28	2 0 0 0 0 0 2 0 0 0 0 0 0 1
90	1,1-Dichloroethylene	180 28	1 0 0 0 0 1 2 0 0 0 0 0 0 1
91	1,2-Dichloroethylene	180 28	0 0 0 0 0 1 2 0 0 0 0 0 0 1
92	Dichloromethane	28 7	1 0 0 0 0 0 2 0 0 0 0 0 0 0
93	2,4-Dichlorophenol	70 7.3	0 0 0 0 1 0 2 0 1 0 0 0 0 0
94	1,3-Dichloropropene	11.3 5.5	2 0 0 0 0 2 2 0 0 0 0 0 0 0
97	Diethyl-phthalate	56 3	2 0 2 2 0 0 0 0 1 0 0 0 0 0
98	Diethylstilbestrol	180 28	4 0 0 2 1 0 0 2 0 0 1 0 1
99	Dihydrosafrole	28 7	1 0 0 0 0 1 0 0 1 0 0 2 1 0
100	Dimethoate	37 11	4 2 0 2 0 1 0 1 0 0 0 0 0 0
101	3,3'-Dimethoxybenzidine	180 28	2 0 0 2 0 0 0 2 2 0 0 0 0 1
102	1,2-Dimethyl-hydrazine	28 7	3 0 0 1 1 0 1 1 2 0 0 1 1 0
103	1,1-Dimethyl-hydrazine	22 8	2 0 0 0 0 0 0 2 0 0 0 0 0 0
104	Dimethyl-terephthalate	28 7	2 0 2 2 0 0 0 0 1 0 0 0 0 0
105	Dimethylamine	14 3.6	2 0 0 0 0 0 1 0 0 0 0 0 0 0
106	Dimethylaminoazobenze	28 7	2 0 0 0 0 1 0 3 2 0 0 0 0 0
107	N,N-Dimethylaniline	180 28	2 0 0 0 0 0 0 1 1 0 0 0 0 1
108	7,12-Dimethylbenz(a)anthrace	28 20	2 0 0 0 0 0 0 0 4 3 2 0 0 0
109	2,4-Dimethylphenol	7 1	2 0 0 0 1 0 0 0 1 0 0 0 0 0
110	Dimetyl-phthalate	7 1	2 0 2 2 0 0 0 0 1 0 0 0 0 0
111	4,6-Dinitro-o-cresol	21 7	1 0 0 0 1 2 0 2 1 0 0 0 0 0
112	1,3-Dinitrobenzene	180 28	0 0 0 0 0 2 0 2 1 0 0 0 0 1

124

TABLE 2. Listing of 258 chemicals in the database, with the low and high values of the half-lives in days, and the input strings used to represent each chemical in training the neural network (continued).

113 2,4-Dinitrophenol	263 68		0 0 0 0 1 0 0 2 1 0 0 0 0 1
114 2,4-Dinitrotoluen	180 28		1 0 0 0 0 0 0 2 1 0 0 0 0 1
115 2,5-Dinitrotoluene	180 28		1 0 0 0 0 2 0 2 1 0 0 0 0 1
116 2,3-Dinitrotoluene	180 28		1 0 0 0 0 2 0 2 1 0 0 0 0 1
117 2,6-Dinitrotoluene	180 28		1 0 0 0 0 2 0 2 1 0 0 0 0 1
118 3,4-Dinitrotoluene	180 28		1 0 0 0 0 2 0 2 1 0 0 0 0 1
119 Dinoseb	123 43		3 0 0 0 1 2 0 2 1 0 0 0 0 0
120 1,4-Dioxane	180 28		0 0 0 0 0 0 0 0 0 0 0 2 1 1
121 Diphenylamine	28 7		0 0 0 0 0 0 0 1 2 0 0 0 0 0
122 Disulfoton	21 3		2 0 1 2 0 1 0 3 0 0 0 0 0 0
123 Epichlorohydrin	28 7		1 0 0 0 0 0 1 0 0 0 0 0 1 0
124 2-Ethoxyethanol	28 7		2 0 0 1 1 0 0 0 0 0 0 0 0 0
125 Ethyl-N-methyl-N-nitrosocarbamate	1 0.5		2 0 1 1 0 0 0 1 0 0 0 0 0 0
126 Ethyl-acetate	7 1		2 0 1 1 0 0 0 0 0 0 0 0 0 0
127 Ethyl-acrylate	7 1		1 0 1 1 0 1 0 0 0 0 0 0 0 0
128 Ethyl-carbamate	7 1		1 0 1 1 0 0 0 1 0 0 0 0 0 0
129 Ethyl-carbethoxymethyl-phthalate		28 7	2 0 3 3 0 0 0 0 1 0 0 0 3 0
130 Ethylbenzene	10 3		1 0 0 0 0 0 0 0 1 0 0 0 0 0
131 Ethylene	28 1		1 0 0 0 0 1 0 0 0 0 0 0 0 0
132 Ethylene-dibromide	180 28		2 0.0 0 0 0 2 0 0 0 0 0 0 0 1
133 Ethylene-glycol	12 2		2 0 0 0.2 0 0 0 0 0 0 0 0 0 0
134 Ethylene-oxide	11.9 10.5		0 0 0 0 0 0 0 0 0 0 0 1 0 0
135 Ethylenethiourea	28 7		0 1 0 0 0 0 2 0 0 0 2 1 0
136 Flourene	60 32		0 0 0 0 0 2 0 0 2 0 0 0 1 0
137 Fluorathene	440 140		0 0 0 0 0 0 0 0 3 1 0 0 1 1
138 Fluridone	192 44		1 1 0 0 0 0 3 1 3 0 0 1 0 1
139 Formaldehyde	7 1		0 1 0 0 0 0 0 0 0 0 0 0 0 0
140 Formic-acid	7 1		1 0 1 0 0 0 1 0 0 0 0 0 0 0
141 Fumazone	180 28		3 0 0 0 0 0 3 0 0 0 0 0 0 1
142 Furan	28 7		0 0 0 0 2 0 0 0 0 0 1 1 0
143 Glycidylaldehyde	28 7		0 1 0 0 0 0 0 0 0 0 0 1 1 0
144 Heptachlor-epoxide	552 33		0 0 0 0 0 1 7 0 0 0 0 1 4 1
145 Hexachlorobenzene	2089 969		0 0 0 0 0 0 6 0 1 0 0 0 0 1
146 Hexachlorobutadiene	180 28		0 0 0 0 0 2 6 0 0 0 0 0 0 1
147 alpha-Hexachlorocyclohexane	240 14		0 0 0 0 0 0 6 0 0 0 0 0 1 1
148 gamma-Hexachlorocyclohexane	135 13.8		0 0 0 0 0 0 6 0 0 0 0 0 1 1
149 Hexachlorocyclopentadiene	28 7		0 0 0 0 0 2 6 0 0 0 0 0 1 0
150 Hexachloroethane	180 28		1 0 0 0 0 0 6 0 0 0 0 0 0 1
151 Hexachlorophene	328 250		1 0 0 0 2 0 6 0 2 0 0 0 0 1
152 Hydrazobenzene	180 28		0 0 0 0 0 0 2 2 0 0 0 0 0 1
153 Hydrocyanic-acid	180 28		1 0 0 0 0 2 0 0 0 0 0 0 0 1
154 Hydroquinone	7 1		0 0 0 0 2 0 0 0 1 0 0 0 0 0
155 Indeno(1,2,3-cd)pyrene	730 600		0 0 0 0 0 2 0 0 5 5 2 0 1 1
156 Isobutyl-alcohol	28 7		3 0 0 0 1 0 0 0 0 0 0 0 0 0
157 Isobutyraldehyde	7 1		3 1 0 0 0 0 0 0 0 0 0 0 0 0
158 Isophorone	28 7		3 1 0 0 0 1 0 0 0 0 0 0 1 0
159 Isoprene	28 7		4 0 0 0 0 2 0 0 0 0 0 0 1 0
160 Isopropailn	105 17		4 0 0 0 0 2 0 3 1 0 0 0 0 1
161 Isopropanol	7 1		2 0 0 0 1 0 0 0 0 0 0 0 0 0
162 4,4'-Isopropylidenediphenol	180 1		2 0 0 0 2 0 0 0 2 0 0 0 0 1
163 Isosafrole	28 7		1 0 0 0 0 1 0 0 1 0 0 2 1 0
164 Kepone	730 312		0 1 0 0 0 0 1 0 0 0 0 0 4 0 1
165 Lasiocarpine	28 7		6 0 2 3 2 2 0 1 0 0 0 1 2 0
166 Linuron	178 28		2 1 0 1 0 0 2 2 1 0 0 0 0 1

TABLE 2. Listing of 258 chemicals in the database, with the low and high values of the half-lives in days, and the input strings used to represent each chemical in training the neural network (continued).

167 Malathion	7 3	4 0 3 4 0 1 0 0 0 0 0 0 0 0
168 Mecoprop	10 7	2 0 1 2 0 0 1 0 1 0 0 0 0 0
169 Melamine	180 28	0 0 0 0 0 3 0 6 1 0 0 3 0 1
170 Methanol	7 1	1 0 0 0 1 0 0 0 0 0 0 0 0 0
171 Methoxychlor	365 180	2 0 0 2 0 0 3 0 2 0 0 0 0 1
172 2-Methoxyethanol	28 7	3 0 0 1 2 0 0 0 0 0 0 0 0 0
173 2-Methyl4chlorophenoxyacetic	7 4	2 0 1 1 0 0 1 0 1 0 0 0 0 0
174 Methyl-acrylate	7 1	1 0 1 1 0 1 0 0 0 0 0 0 0 0
175 Methyl-bromide	28 7	1 0 0 0 0 0 1 0 0 0 0 0 0 0
176 Methyl-chloride	28 7	1 0 0 0 0 0 1 0 0 0 0 0 0 0
177 Methyl-ethyl-ketone	7 1	2 1 0 0 0 0 0 0 0 0 0 0 0 0
178 Methyl-ethyl-ketone-peroxide	365 180	2 1 0 0 1 1 0 1 0 0 0 0 0 1
179 Methyl-iodide	28 7	1 0 0 0 0 2 0 0 0 0 0 0 0 0
180 Methyl-isobutyl-ketone	7 1	5 1 0 0 0 0 0 0 0 0 0 0 0 0
181 Methyl-methacrylate	28 7	2 0 1 1 0 1 0 0 0 0 0 0 0 0
182 Methyl-parathion	365 10	2 0 1 3 0 1 0 1 1 0 0 0 0 1
183 2-Methylaziridine	36 3.6	1 0 0 0 0 0 1 0 0 0 1 0 0
184 3-Methylcholanthrene	1400 609	1 0 0 0 0 0 0 4 3 2 0 1 1
185 Methylene-bromide	28 7	1 0 0 0 0 2 0 0 0 0 0 0 0 0
186 Methylhydrazine	24 13	1 0 0 0 0 2 0 0 0 0 0 0
187 Methylthiouracil	28 7	1 2 0 0 0 0 2 0 0 0 1 0 0
188 Michlers-ketone	28 7	4 1 0 0 0 0 2 2 0 0 0 0 0
189 Mitomycin-C	28 7	2 3 1 1 0 0 0 4 1 0 0 1 2 0
190 N-Nitrosodiphenylamione	34 10	0 0 0 0 0 1 2 2 0 0 0 0 0
191 N-Nitrosopiperidine	180 28	0 0 0 0 1 0 2 0 0 0 1 0 1
192 N-Nitrosopyrrolidine	180 28	0 0 0 0 1 0 2 0 0 0 1 1 1
193 Naphthalene	48 16.6	0 0 0 0 0 0 0 1 1 0 0 0 0
195 beta-Naphthylamine	180 28	0 0 0 0 0 0 1 2 1 0 0 0 1
196 Nitrilotriacetic-acid	28 3	3 0 3 0 0 0 0 1 0 0 0 0 0 0
197 5-Nitro-o-anisidine	28 1	1 0 0 1 0 1 0 2 1 0 0 0 0 0
198 5-Nitro-o-toluidine	28 1	1 0 0 0 0 1 0 2 1 0 0 0 0 0
199 Nitrobenzene	197 13.4	0 0 0 0 1 0 1 1 0 0 0 0 1
200 4-Nitrobiphenyl	28 1	0 0 0 0 0 1 0 1 2 0 0 0 0 0
201 Nitroglycerin	7 2	3 0 0 0 0 3 0 3 0 0 0 0 0 0
202 2-Nitrophenol	28 7	0 0 0 0 1 1 0 1 1 0 0 0 0 0
203 2-Nitropropane	180 28	3 0 0 0 0 1 0 1 0 0 0 0 0 1
204 N-Nitrosodiethylamine	180 20	0 0 0 0 0 1 0 2 2 0 0 0 0 1
205 N-Nitrosodimethylamine	180 21	1 0 0 0 0 1 0 1 2 0 0 0 0 1
206 p-Nitrosodiphenylamine	180 28	0 0 0 0 0 1 0 2 2 0 0 0 0 1
207 N-Nitrosomorpholine	180 28	0 0 0 0 0 1 0 2 0 0 0 2 1 1
208 Octrachloronaphthalene	365 180	0 0 0 0 0 0 8 2 1 0 0 0 0 1
209 Pentachlorobenzene	345 194	0 0 0 0 0 0 5 0 1 0 0 0 0 1
210 Pentachloronitrobenzene	699 213	0 0 0 0 0 1 5 1 1 0 0 0 0 1
211 Petachlorophenol	178 23	0 0 0 0 1 0 5 0 1 0 0 0 0 1
212 Phenacetin	28 7	2 1 0 1 0 0 0 1 1 0 0 0 0 0
213 Phenanthrene	200 16	0 0 0 0 0 0 0 3 2 3 0 0 1
214 2-Phenlphenol	7 1	0 0 0 1 0 0 0 2 0 0 0 0 0
215 Phenobarbital	28 7	1 3 0 0 0 0 2 1 0 0 1 0 0
216 Phenol	10 1	0 0 0 0 1 0 0 0 1 0 0 0 0 0
217 p-Phenylenediamine	28 7	0 0 0 0 0 0 0 2 1 0 0 0 0 0
218 Picric-acid	180 28	0 0 0 0 1 3 0 3 1 0 0 0 0 1
219 Propionadehyde	7 1	1 1 0 0 0 0 0 0 0 0 0 0 0 0
220 Propoxur	28 1.6	3 0 1 2 0 0 0 1 1 0 0 0 0 0
221 Propylene	28 7	1 0 0 0 0 1 0 0 0 0 0 0 0 0

TABLE 2.   Listing of 258 chemicals in the database, with the low and high values of the half-lives in days, and the input strings used to represent each chemical in training the neural network (continued).

222 Propylene-glycol,monoethyl-ether	28 7	2 0 0 1 1 0 0 0 0 0 0 0 0 0
223 Propylene-glycol,monomethyl-ether	28 7	2 0 0 1 1 0 0 0 0 0 0 0 0 0
224 Pyrene	1900 210	0 0 0 0 0 0 0 0 4 5 2 0 0 1
225 Quinoline	10 3	0 0 0 0 0 0 0 1 2 1 0 1 0 0
226 Saccharin	28 7	0 1 0 0 0 2 0 1 1 0 0 2 1 0
227 Safrole	28 7	2 0 0 0 0 2 0 0 1 0 0 2 1 0
228 Streptozotocin	28 7	5 2 0 0 4 2 0 2 0 0 0 0 0 0
229 Strychnine	28 7	0 1 0 0 0 0 0 2 1 0 0 4 1 0
230 Styrene	28 14	0 0 0 0 0 1 0 0 1 0 0 0 0 0
231 1,2,4,5-Tetrachlorobenzene	180 28	0 0 0 0 0 0 4 0 1 0 0 0 0 1
232 1,1,2,2-Tetrachloroethane	45 0.5	2 0 0 0 0 0 4 0 0 0 0 0 0 0
233 Tetrachloroethylene	365 180	0 0 0 0 1 0 4 0 0 0 0 0 0 1
234 2,3,4-Tetrachlorophenol	180 28	0 0 0 0 1 0 4 0 1 0 0 0 0 1
235 Tetraethyl-lead	28 7	4 0 0 0 0 0 0 0 0 0 0 0 0 0
236 Thioacetamide	7 1	2 1 0 0 0 0 0 1 0 0 0 0 0 0
237 4,4-Thiodianiline	28 7	0 0 0 0 0 0 3 2 0 0 0 0 0 0
238 Thiourea	7 1	1 1 0 0 0 0 2 0 0 0 0 0 0 0
239 Toluen	22 4	1 0 0 0 0 0 0 0 1 0 0 0 0 0
240 o-Toluidine	7 1	1 0 0 0 0 0 1 1 0 0 0 0 0 0
241 Triaziquone	31 7	0 2 0 0 0 0 3 1 0 0 3 0 0 0
242 Trichloro-1,2,2-trifluoroethane	365 180	1 0 0 0 0 0 6 0 0 0 0 0 0 1
243 1,2,4-Trichlorobenzene	108 28	0 0 0 0 0 0 3 0 1 0 0 0 0 1
244 1,1,1-Trichloroethane	273 140	2 0 0 0 0 0 3 0 0 0 0 0 0 1
245 1,1,2-Trichloroethane	365 134	2 0 0 0 0 0 3 0 0 0 0 0 0 1
246 Trichloroethylene	365 180	0 0 0 0 0 1 3 0 0 0 0 0 0 1
247 Trichlorofluoromethane	365 180	1 0 0 0 0 0 4 0 0 0 0 0 0 1
248 Trichlorofon	45 1	0 0 1 2 1 0 3 0 0 0 0 0 0 0
249 2,4,6-Trichlorophenol	70 7	0 0 0 0 1 0 3 0 1 0 0 0 0 1
250 2,4,5-Trichlorophenoxyacetic	20 10	1 0 1 1 0 0 3 0 1 0 0 0 0 0
255 m-Xylene	28 7	2 0 0 0 0 0 0 0 1 0 0 0 0 0
256 o-Xylene	28 7	2 0 0 0 0 0 0 0 1 0 0 0 0 0
257 p-Xylene	28 7	2 0 0 0 0 0 0 0 1 0 0 0 0 0
258 2,6-Xylidine	316 3	2 0 0 0 0 0 1 1 0 0 0 0 0 1
16 4-Aminoazobenzene	28 7	0 0 0 0 0 1 0 3 2 0 0 0 0 0
17 4-Aminobiphenyl	7 1	0 0 0 0 0 0 1 2 0 0 0 0 0 0
18 Amitrole	180 28	0 0 0 0 0 2 0 4 0 0 0 3 1 1
95 2,4-Dichlorpheoxyacetic-acid	50 10	1 0 1 1 0 0 2 0 1 0 0 0 0 0
96 Dieldrin	1080 175	0 0 0 0 0 1 6 0 0 0 0 1 4 1
194 alpha-Naphthylamine	180 28	0 0 0 0 0 0 0 1 2 1 0 0 0 1
251 1,2,3-Trichloropropane	365 180	3 0 0 0 0 0 3 0 0 0 0 0 0 1
252 1,2,4-Trimethylbenzene	28 7	3 0 0 0 0 0 0 0 1 0 0 0 0 0
253 Vinyl-chloride	180 28	1 0 0 0 0 1 1 0 0 0 0 0 0 1
254 Warfarin	28 7	1 2 1 1 1 1 0 0 2 0 0 1 1 0

network to estimate the low and high values for the half-lives. The error for the testing compounds were within the same range, and hence results are only reported here for a specified testing set, wherein the chemicals represent the variety of chemicals in the database, as mentioned earlier.

5. Results and Discussion

The input string database, listed in Table 2, were used to train and test the neural network using the back-propagation method. The neural network parameters are listed in Table 3. The estimation results for all compounds are shown in Table 4. The estimation error, E_i, expressed as a percent, is defined as follows:

$$E_i(\%) = \frac{(t_{1/2,NN} - t_{1/2,LIT})}{t_{1/2,LIT}} * 100 \qquad (3)$$

where $t_{1/2,NN}$ is the half-life predicted by the neural network and $t_{1/2,LIT}$ is the value of the half-life reported in the literature.

TABLE 3.   Listing of the neural network parameters.

Parameter	A	B
number of compounds in the training set	248	248
number of predicted compounds	10	10
input units	14	13
hidden units	12	12
output units	2	2
maximum-epoch-number[a]	200	200
learning step, η	0.35	0.35
momentum, $\alpha 0$	0.001[b]	0.001
normal factor[c]	9	9

a. Training stops when iteration/epoch number reaches this specified maximum-epoch-number.

b. The momentum is set to zero at the first 10 epoches and then to $\alpha 0$, i.e.:

  $\alpha = 0.0$, if number of iterations/epoches < 10;

  $\alpha = \alpha 0$, otherwise.

c. The desired output value are normalized via
  ln(desired half-life)/norm_factor.

More than 80 % of the chemicals in the selected training set have a relative error less than 50 % and most of them are less than 20 %, as shown in Figure 2. Most of the points are located near the 45" line. Possible errors may be due to many reasons, which

128

includes: (1) errors in the reported half-lives; and (2) the chosen indicators do not adequately describe the chemical.

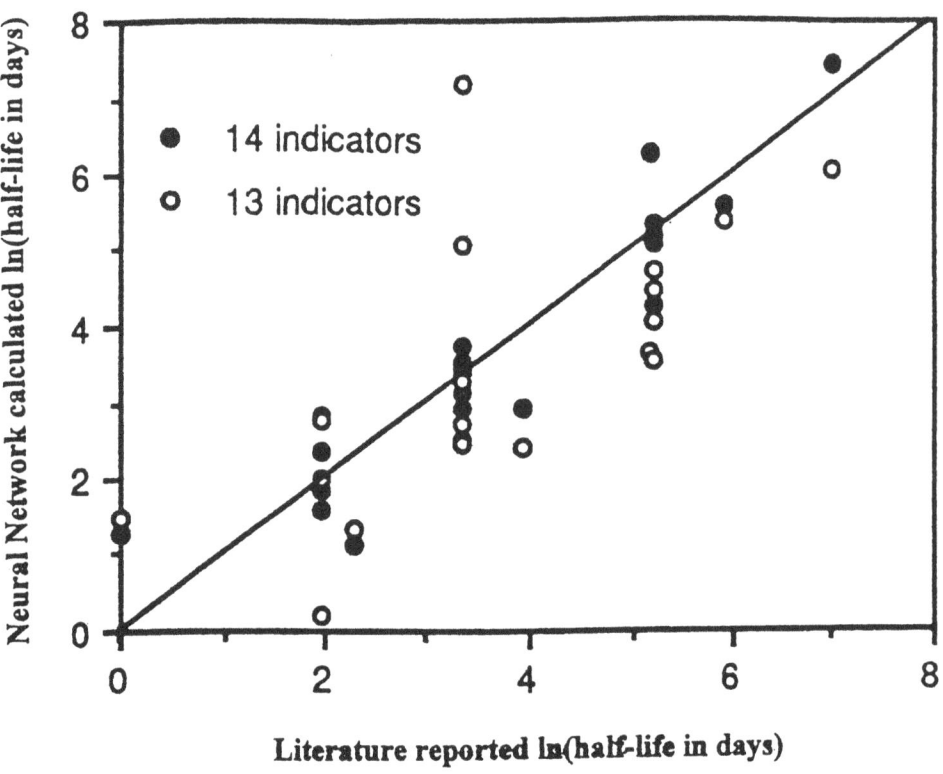

Figure 2. Plot of literature reported ln(half-life in days) versus neural network calculated value for ln(half-life in days) for each chemical. Both the high and low values have been plotted for each chemical.

TABLE 4. Comparison of the half-lives reported in the literature and calculated by the neural network. Note: exp[a] is the unacclimated soil half-life ($t_{1/2,LIT}$) reported in the literature. est[b] is the half-life value ($t_{1/2,NN}$) estimated using the neural network. err[c] (%) is the percent error defined by equation (3).

| Chemical | High | | | Low | | |
	exp[a]	est[b]	err[c](%)	exp	est.	err(%)
1 Acenaphylene	60.0	96.5	61	42.5	40.3	-5
2 Acetamide	7.0	8.1	16	1.0	1.4	38
3 Acetone	7.0	6.8	-3	1.0	1.4	42
4 Acetonitrile	28.0	18.7	-33	7.0	3.9	-44
5 2-Acetylaminofluorene	180.0	145.1	-19	28.0	24.5	-12
6 Acrolein	28.0	15.5	-45	7.0	3.0	-56
7 Acrylic acid	7.0	9.7	39	1.0	1.7	67
8 Acrylonitrile	23.0	19.2	-17	1.3	2.6	104
9 Aflatoxin B1	28.0	33.8	21	7.0	6.0	-15
10 Aldicarb	361.0	416.9	15	20.0	28.0	40
11 Alfrin	592.0	237.2	-60	21.0	30.0	43
12 Allyl alcohol	7.0	10.3	47	1.0	1.9	86
13 1-Amino-2-methylanthraquinone	28.0	23.8	-15	7.0	5.6	-20
14 2-Aminoanthraquinone	28.0	43.7	56	7.0	8.1	16
15 o-Anisidine	180.0	161.4	-10	28.0	19.1	-32
16 p-Anisidine	28.0	25.9	-8	7.0	5.7	-19
17 Anthracene	460.0	406.5	-12	50.0	47.7	-5
18 Auramine	180.0	271.6	51	28.0	27.3	-3
19 Azaserine	56.0	30.0	-47	8.2	7.2	-13
20 Aziridine	28.0	32.9	17	7.0	8.2	17
21 Benefin	120.0	86.5	-28	21.0	22.1	5
22 Benz(a)anthracene	680.0	799.9	18	102.0	193.5	90
23 Benz(c)acridine	365.0	547.8	50	180.0	151.4	-16
24 Benzamide	15.0	23.2	55	2.0	2.9	47
25 Benzene	16.0	72.5	353	5.0	6.3	26
26 1,3-Benzenediamine	28.0	29.1	4	7.0	6.4	-9
27 Benzidine	8.0	20.1	151	2.0	4.9	145
28 Benzo(a)pyrene	530.0	715.9	35	57.0	187.5	229
29 Benzo(b)fluoranthene	610.0	612.7	0	360.0	282.0	-22
30 Benzo(ghi)perylene	650.0	535.7	-18	590.0	430.4	-27
31 Benzo(k)fluranthene	2140.0	1787.5	-16	910.0	1057.0	16
32 Biphenyl	7.0	6.8	-2	1.5	2.1	38
33 Bis(2-choloethyl)ether	180.0	221.4	23	28.0	28.7	3
34 Bromoethylene	180.0	164.2	-9	28.0	34.4	23
35 Bromoform	180.0	206.0	14	28.0	52.3	87
36 1,3-Butadiene	28.0	26.8	-4	7.0	3.6	-49
37 Butanol	7.0	9.9	41	1.0	1.6	62
38 sec-Butyl alcohol	7.0	8.9	28	1.0	1.5	54
39 tert-Butyl alcohol	200.0	214.8	7	15.0	11.6	-23

TABLE 4. Comparison of the half-lives reported in the literature and calculated by the neural network. Note: exp[a] is the unacclimated soil half-life ($t_{1/2,LIT}$) reported in the literature. est[b] is the half-life value ($t_{1/2,NN}$) estimated using the neural network. err[c] (%) is the percent error defined by equation (3) (continued).

	Half-lives (days)					
		High			Low	
Chemical	exp[a]	est[b]	err[c](%)	exp	est.	err(%)
40 Butyl-benzyl phthalate	7.0	7.1	1	1.0	1.5	52
41 1,2-Butylene oxide	12.9	14.6	13	7.0	6.0	-15
42 Butylglycolyl butyl phthalate	7.0	9.4	35	1.0	1.4	36
43 C.I. Basic Green 4	180.0	167.7	-7	28.0	35.2	26
44 C.I. Solvent Yellow 3	28.0	18.7	-33	7.0	3.6	-48
45 C.I. Solvent Yellow 4	28.0	36.0	29	7.0	5.8	-17
46 C.I. Vat Yellow 4	180.0	145.1	-19	28.0	34.1	22
47 Captan	60.0	27.5	-54	2.0	2.9	46
48 Carbon-tetrachloride	265.0	219.0	-17	180.0	53.7	-70
49 Catechol	7.0	4.1	-41	1.0	1.0	2
50 Chlordane	1386.0	1816.3	31	238.0	243.7	2
51 Chloroacetic acid	7.0	9.1	29	1.0	2.0	95
52 2-Chloroacetophenone	28.0	25.8	-8	7.0	4.6	-34
53 Chlorobenzene	150.0	104.5	-30	68.0	14.2	-79
54 Chlorobenzilate	35.0	39.1	12	7.0	8.1	16
55 Chloroethane	28.0	27.1	-3	7.0	6.4	-9
56 Chloroform	180.0	220.2	22	28.0	62.3	122
57 Chloroprene	180.0	157.9	-12	28.0	33.6	20
58 3-Chloropropene	14.0	28.3	102	6.9	5.7	-17
59 Chrysene	1000.0	1022.9	2	371.0	260.5	-30
60 p-Cresidine	180.0	226.8	26	28.0	44.3	58
61 o-Cresol	7.0	6.2	-11	1.0	1.2	21
62 Crotonaldehyde (trans)	7.0	9.6	38	1.0	2.5	146
63 Cumene	8.0	16.4	104	2.0	4.5	124
64 Cumene hydroperoxide	28.0	39.3	40	7.0	7.0	1
65 Cupferron	180.0	153.6	-15	28.0	31.7	13
66 Cyclophosphamide	28.0	48.4	73	7.0	5.1	-28
67 2,4-DB	7.0	6.7	-4	1.0	1.7	72
68 DDD	5833.0	2383.1	-59	730.0	806.9	11
69 DDT	5833.0	2872.8	-51	730.0	1110.5	52
70 Dalapon	60.0	33.8	-44	14.0	11.8	-15
71 Di-n-butyl-phthalate	23.0	22.1	-4	2.0	2.1	3
72 Diallate	90.0	49.4	-45	10.5	13.4	28
73 2,4-Diaminotoluene	180.0	149.2	-17	28.0	22.6	-19
74 Diaminotoluenes	180.0	152.1	-16	28.0	23.1	-17
75 Dibenz(a,h)anthracene	940.0	525.7	-44	361.0	102.7	-72
76 Dibenzofyran	28.0	23.6	-16	7.0	6.0	-15
77 1,2,7,8-Dibenzopyrene	361.0	621.8	72	232.0	132.5	-43
78 Dibromochloromethane	180.0	218.2	21	28.0	60.7	117

TABLE 4. Comparison of the half-lives reported in the literature and calculated by the neural network. Note: exp[a] is the unacclimated soil half-life ($t_{1/2.LIT}$) reported in the literature. est[b] is the half-life value ($t_{1/2.NN}$) estimated using the neural network. err[c] (%) is the percent error defined by equation (3) (continued).

	Half-lives (days)					
		High			Low	
Chemical	exp[a]	est[b]	err[c](%)	exp	est.	err(%)
79 m-Dichlorobenzene	180.0	201.6	12	28.0	30.4	9
80 1,2-Dichlorobenzene	180.0	199.0	11	28.0	30.0	7
81 p-Dichlorobenzene	180.0	196.6	9	28.0	29.7	6
82 3,3'-Dichlorobenzidine	180.0	200.9	12	28.0	44.7	60
83 Dichlorodifluoromethane	180.0	216.1	20	28.0	57.1	104
84 1,1-Dichloroethane	154.0	32.9	-79	32.0	5.6	-82
85 1,2-Dichloroethane	180.0	223.5	24	28.0	61.9	121
86 1,1-Dichloroethylene	180.0	212.8	18	28.0	44.3	58
87 1,2-Dichloroethylene	180.0	220.9	23	28.0	42.7	52
88 Dichloromethane	28.0	52.6	88	7.0	11.2	60
89 2,4-Dichlorophenol	70.0	56.1	-20	7.3	6.5	-11
90 1,3-Dichloropropene	11.3	26.9	138	5.5	3.3	-40
91 Diethyl-phthalate	56.0	13.6	-76	3.0	1.6	-47
92 Diethylstilbestrol	180.0	233.5	30	28.0	19.7	-30
93 Dihydrosafrole	28.0	25.0	-11	7.0	8.2	18
94 Dimethoate	37.0	37.4	1	11.0	9.8	-11
95 3,3'-Dimethoxybenzidine	180.0	146.2	-19	28.0	22.7	-19
96 1,2-Dimethyl-hydrazine	28.0	24.2	-14	7.0	7.3	4
97 1,1-Dimethyl-hydrazine	22.0	27.7	26	8.0	8.1	2
98 Dimethyl-terephthalate	28.0	22.4	-20	7.0	1.8	-75
99 Dimethylamine	14.0	15.8	13	3.6	3.6	1
100 Dimethylaminoazobenze	28.0	23.4	-17	7.0	4.0	-44
101 N,N-Dimethylaniline	180.0	137.0	-24	28.0	9.7	-66
102 7,12-Dimethylbenz(a)anthrace	28.0	25.9	-7	20.0	20.6	3
103 2,4-Dimethylphenol	7.0	13.1	88	1.0	1.5	53
104 Dimetyl phthalate	7.0	25.3	261	1.0	1.9	94
105 4,6-Dinitro-o-cresol	21.0	22.1	5	7.0	4.6	-34
106 1,3-Dinitrobenzene	180.0	172.1	-4	28.0	32.0	14
107 2,4-Dinitrophenol	263.0	238.8	-9	68.0	68.5	1
108 2,4-Dinitrotoluen	180.0	154.6	-14	28.0	28.3	1
109 2,5-Dinitrotoluene	180.0	174.7	-3	28.0	24.3	-13
110 2,3-Dinitrotoluene	180.0	175.4	-3	28.0	24.6	-12
111 2,6-Dinitrotoluene	180.0	176.1	-2	28.0	24.9	-11
112 3,4-Dinitrotoluene	180.0	176.6	-2	28.0	25.2	-10
113 Dinoseb	123.0	68.6	-44	43.0	72.5	69
114 1,4-Dioxane	180.0	142.3	-21	28.0	29.4	5
115 Diphenylamine	28.0	13.3	-52	7.0	3.5	-51
116 Disulfoton	21.0	18.6	-12	3.0	3.0	-1
117 Epichlorohydrin	28.0	29.4	5	7.0	7.6	8

TABLE 4. Comparison of the half-lives reported in the literature and calculated by the neural network. Note: exp[a] is the unacclimated soil half-life ($t_{1/2,LIT}$) reported in the literature. est[b] is the half-life value ($t_{1/2,NN}$) estimated using the neural network. err[c] (%) is the percent error defined by equation (3) (continued).

	Half-lives (days)					
	High				Low	
Chemical	exp[a]	est[b]	err[c](%)	exp	est.	err(%)
118 2-Ethoxyethanol	28.0	27.4	-2	7.0	7.7	10
119 Ethyl-N-methyl-N-nitrosocarbamate						
	1.0	3.0	202	0.5	1.0	108
120 Ethyl acetate	7.0	5.7	-18	1.0	1.3	35
121 Ethyl acrylate	7.0	10.3	47	1.0	2.2	118
122 Ethyl carbamate	7.0	3.4	-52	1.0	1.1	10
123 Ethyl carbethoxymethyl phthalate						
	28.0	28.1	0	7.0	7.0	0
124 Ethylbenzene	10.0	13.4	34	3.0	3.4	12
125 Ethylene	28.0	16.5	-41	1.0	3.7	271
126 Ethylene dibromide	180.0	214.6	19	28.0	50.7	81
127 Ethylene glycol	12.0	15.1	26	2.0	2.1	3
128 Ethylene oxide	11.9	17.0	43	10.5	4.5	-57
129 Ethylenethiourea	28.0	24.4	-13	7.0	5.0	-28
130 Flourene	60.0	42.6	-29	32.0	14.1	-56
131 Fluorathene	440.0	474.7	8	140.0	111.0	-21
132 Fluridone	192.0	225.7	18	44.0	46.9	7
133 Formaldehyde	7.0	6.8	-3	1.0	1.3	34
134 Formic acid	7.0	9.8	39	1.0	2.1	106
135 Fumazone	180.0	222.5	24	28.0	55.2	97
136 Furan	28.0	27.6	-1	7.0	5.0	-28
137 Glycidylaldehyde	28.0	22.7	-19	7.0	9.1	31
138 Heptachlor-epoxide	552.0	513.2	-7	33.0	34.4	4
139 Hexachlorobenzene	2089.0	611.9	-71	969.0	104.5	-89
140 Hexachlorobutadiene	180.0	269.5	50	28.0	57.9	107
141 alpha-Hexachlorocyclohexane	240.0	141.7	-41	14.0	15.8	13
142 gamma-Hexachlorocyclohexane						
	135.0	149.9	11	13.8	15.9	15
143 Hexachlorocyclopentadiene	28.0	50.4	80	7.0	4.1	-41
144 Hexachloroethane	180.0	241.7	34	28.0	61.8	121
145 Hexachlorophene	328.0	388.6	18	250.0	227.5	-9
146 Hydrazobenzene	180.0	128.7	-29	28.0	18.7	-33
147 Hydrocyanic-acid	180.0	208.0	16	28.0	20.5	-27
148 Hydroquinone	7.0	4.1	-42	1.0	1.0	2
149 Indeno(1,2,3-cd)pyrene	730.0	1059.3	45	600.0	841.7	40
150 Isobutyl-alcohol	28.0	13.6	-52	7.0	2.2	-68
151 Isobutyraldehyde	7.0	7.4	6	1.0	1.5	49
152 Isophorone	28.0	20.3	-28	7.0	6.1	-12
153 Isoprene	28.0	44.0	57	7.0	8.3	18
154 Isopropailn	105.0	241.9	130	17.0	19.6	15
155 Isopropanol	7.0	9.5	35	1.0	1.7	66
156 4,4'-Isopropylidenediphenol	180.0	126.8	-30	1.0	1.9	88

TABLE 4. Comparison of the half-lives reported in the literature and calculated by the neural network. Note: exp[a] is the unacclimated soil half-life ($t_{1/2,LIT}$) reported in the literature. est[b] is the half-life value ($t_{1/2,NN}$) estimated using the neural network. err[c] (%) is the percent error defined by equation (3) (continued).

	Half-lives (days)					
		High			Low	
Chemical	exp[a]	est[b]	err[c](%)	exp	est.	err(%)
157 Isosafrole	28.0	25.9	-8	7.0	9.1	30
158 Kepone	730.0	766.7	5	312.0	275.7	-12
159 Lasiocarpine	28.0	27.0	-4	7.0	6.9	-1
160 Linuron	178.0	204.1	15	28.0	24.1	-14
161 Malathion	7.0	6.9	-2	3.0	3.4	13
162 Mecoprop	10.0	12.4	24	7.0	3.3	-53
163 Melamine	180.0	183.4	2	28.0	32.3	15
164 Methanol	7.0	4.9	-30	1.0	1.1	15
165 Methoxychlor	365.0	264.4	-28	180.0	231.9	29
166 2-Methoxyethanol	28.0	26.6	-5	7.0	7.9	12
167 2-Methyl-4-chlorophenoxyacetic acid						
	7.0	7.1	1	4.0	1.6	-59
168 Methyl acrylate	7.0	9.5	36	1.0	2.0	103
169 Methyl bromide	28.0	27.1	-3	7.0	5.8	-17
170 Methyl chloride	28.0	27.3	-3	7.0	5.9	-16
172 Methyl ethyl ketone	7.0	5.6	-20	1.0	1.3	33
172 Methyl ethyl ketone peroxide	365.0	281.7	-23	180.0	166.5	-8
173 Methyl iodide	28.0	52.9	89	7.0	11.1	59
174 Methyl isobutyl ketone	7.0	7.0	-0	1.0	1.5	45
175 Methyl methacrylate	28.0	24.8	-11	7.0	4.4	-37
176 Methyl parathion	365.0	288.9	-21	10.0	10.9	9
177 2-Methylaziridine	36.0	19.8	-45	3.6	5.9	65
178 3-Methylcholanthrene	1400.0	938.6	-33	609.0	724.3	19
179 Methylene-bromide	28.0	54.2	94	7.0	10.0	43
180 Methylhydrazine	24.0	26.8	12	13.0	7.5	-43
181 Methylthiouracil	28.0	31.4	12	7.0	7.9	13
182 Michlers-ketone	28.0	23.2	-17	7.0	5.5	-22
183 Mitomycin C	28.0	27.1	-3	7.0	7.0	-1
184 N-Nitrosodiphenylamione	34.0	36.6	8	10.0	8.3	-17
185 N-Nitrosopiperidine	180.0	175.4	-3	28.0	28.9	3
186 N-Nitrosopyrrolidine	180.0	172.4	-4	28.0	26.8	-4
187 Naphthalene	48.0	43.2	-10	16.6	14.6	-12
188 beta-Naphthylamine	180.0	194.7	8	28.0	27.8	-1
189 Nitrilotriacetic acid	28.0	30.7	10	3.0	2.2	-26
190 5-Nitro-o-anisidine	28.0	19.5	-30	1.0	2.9	186
191 5-Nitro-o-toluidine	28.0	21.4	-23	1.0	3.2	224
192 Nitrobenzene	197.0	185.8	-6	13.4	27.2	103
193 4-Nitrobiphenyl	28.0	25.3	-10	1.0	4.9	390
194 Nitroglycerin	7.0	7.4	5	2.0	1.3	-37
195 2-Nitrophenol	28.0	28.0	-0	7.0	3.8	-45
196 2-Nitropropane	180.0	219.5	22	28.0	12.5	-55
197 N-Nitrosodiethylamine	180.0	199.6	11	20.0	31.0	55

TABLE 4. Comparison of the half-lives reported in the literature and calculated by the neural network. Note: exp[a] is the unacclimated soil half-life ($t_{1/2,LIT}$) reported in the literature. est[b] is the half-life value ($t_{1/2,NN}$) estimated using the neural network. err[c] (%) is the percent error defined by equation (3) (continued).

Chemical	Half-lives (days)					
	High			Low		
	exp[a]	est[b]	err[c](%)	exp	est.	err(%)
198 N-Nitrosodimethylamine	180.0	172.4	-4	21.0	15.9	-24
199 p-Nitrosodiphenylamine	180.0	198.7	10	28.0	30.6	9
200 N-Nitrosomorpholine	180.0	197.1	10	28.0	29.1	4
201 Octachloronaphthalene	365.0	406.5	11	180.0	200.4	11
202 Pentachlorobenzene	345.0	731.7	112	194.0	125.0	-36
203 Pentachloronitrobenzene	699.0	311.5	-55	213.0	68.7	-68
204 Petachlorophenol	178.0	148.4	-17	23.0	17.9	-22
205 Phenacetin	28.0	35.7	27	7.0	7.0	-0
206 Phenanthrene	200.0	216.0	8	16.0	21.2	32
207 2-Phenlphenol	7.0	8.8	26	1.0	1.4	45
208 Phenobarbital	28.0	26.8	-4	7.0	5.2	-26
209 Phenol	10.0	8.0	-20	1.0	1.3	26
210 p-Phenylenediamine	28.0	29.7	6	7.0	6.0	-14
211 Picric acid	180.0	183.3	2	28.0	26.8	-4
212 Propionadehyde	7.0	5.5	-21	1.0	1.3	28
213 Propoxur	28.0	24.9	-11	1.6	2.8	73
214 Propylene	28.0	16.9	-40	7.0	3.4	-51
215 Propylene glycol, monoethyl ether						
	28.0	25.7	-8	7.0	5.6	-21
215 Propylene glycol, monoethyl ether						
	28.0	25.7	-8	7.0	5.6	-21
217 Pyrene	1900.0	1034.9	-46	210.0	330.4	57
218 Quinoline	10.0	9.4	-6	3.0	2.1	-29
219 Saccharin	28.0	44.3	58	7.0	7.1	1
220 Safrole	28.0	33.7	20	7.0	8.1	16
221 Streptozotocin	28.0	35.5	27	7.0	6.1	-14
222 Strychnine	28.0	32.4	16	7.0	9.8	40
223 Styrene	28.0	28.5	2	14.0	7.1	-49
224 1,2,4,5-Tetrachlorobenzene	180.0	660.7	267	28.0	127.7	356
225 1,1,2,2-Tetrachloroethane	45.0	31.0	-31	0.5	2.8	461
226 Tetrachloroethylene	365.0	374.2	3	180.0	211.3	17
227 2,3,4-Tetrachlorophenol	180.0	125.2	-30	28.0	13.1	-53
228 Tetraethyl lead	28.0	26.4	-6	7.0	5.8	-17
229 Thioacetamide	7.0	5.9	-15	1.0	1.2	20
230 4,4-Thiodianiline	28.0	24.6	-12	7.0	5.2	-26
231 Thiourea	7.0	10.3	48	1.0	1.4	42
232 Toluen	22.0	14.0	-36	4.0	3.2	-19
233 o-Toluidine	7.0	18.1	158	1.0	3.5	254
234 Triaziquone	31.0	24.3	-22	7.0	5.7	-18
235 Trichloro-1,2,2-trifluoroethane	365.0	232.0	-36	180.0	50.5	-72
236 1,2,4-Trichlorobenzene	108.0	415.3	285	28.0	65.4	134
237 1,1,1-Trichloroethane	273.0	213.2	-22	140.0	52.9	-62

TABLE 4.   Comparison of the half-lives reported in the literature and calculated by the neural network. Note: exp[a] is the unacclimated soil half-life ($t_{1/2,LIT}$) reported in the literature. est[b] is the half-life value ($t_{1/2,NN}$) estimated using the neural network. err[c] (%) is the percent error defined by equation (3) (continued).

Chemical	Half-lives (days)					
	exp[a]	High est[b]	err[c](%)	exp	Low est.	err(%)
238 1,1,2-Trichloroethane	365.0	223.3	-39	134.0	62.5	-53
239 Trichloroethylene	365.0	236.9	-35	180.0	55.4	-69
240 Trichlorofluoromethane	365.0	242.7	-34	180.0	72.7	-60
241 Trichlorofon	45.0	32.1	-29	1.0	1.5	55
242 2,4,6-Trichlorophenol	70.0	131.7	88	7.0	17.3	146
243 2,4,5-Trichlorophenoxyacetic	20.0	29.2	46	10.0	5.5	-45
244 m-Xylene	28.0	22.0	-21	7.0	6.1	-13
245 o-Xylene	28.0	23.2	-17	7.0	6.4	-9
246 p-Xylene	28.0	24.2	-14	7.0	6.6	-6
247 2,6-Xylidine	316.0	145.6	-54	3.0	12.0	301
248 Acenaphthene	102.0	89.2	-13	12.3	18.4	49

136

The trained neural network is also used to predict the half-lives for ten compounds which are not used to train the neural network. These compounds have not been seen by the network before and the objective of this analysis is to test the accuracy of the predictions. The 10 test set compounds were selected such that all compounds are different, involve all 14 indicators and differ widely in their values of the half-lives. Table 5 shows the results of the predictions made by the neural network. 70 % of the chemicals have a prediction error of less than 50 %. In fact, 35 % gave prediction errors less than 20 %. Only one chemical gave an error greater than 100 %.

To examine the importance of Indicator #14, which is a binary variable indicating whether the chemical structure is symmetric (value = 1) or not (value = 0), the neural network was trained first using all 14 indicators and then again without using Indicator #14. Figure 3 shows the results, wherein the total squared error reduced from 3.4 to 1.4 when Indicator #14 was used in the input, and half-life estimates for 88 out of the 100 symmetric structures present in the database were improved. This shows that symmetrical molecular structures exhibit longer half-lives than non-symmetrical molecules. Although the detailed mechanism for this effect is unclear at this time, analysis of the reported data on soil half-lives for the compounds in the database consistently revealed that symmetrical molecules have slower degradation rates in soil than non-symmetrical molecules.

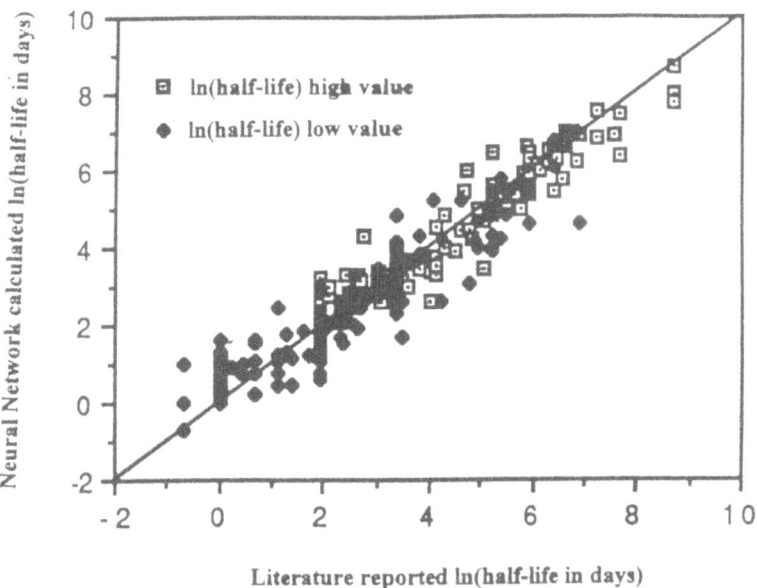

Figure 3. Comparison of the half-life prediction results from the neural network with the literature reported values using 13 and 14 indicators.

The number of nodes in the hidden layer were also varied from 7 to 13 and results were found to be very similar for both the training and testing sets. Further, the selected chemicals can be classified based on the reported data on half-lives: (1) Fast if the low to high range of half-lives are from 1 to 7 days; (2) Moderately fast if the half-life range is from 7 to 28 days; (3) Slow if the half-life range is from 28 days to 180 days; and (4) Resistant if the half-life range is from 180 days to 365 days. Using this classification scheme, the neural network was able to correctly classify more than 95 % of the 258 chemicals in the database.

TABLE 5.   Comparison of literature reported half-lives with half-life values predicted by the neural network for the test set. Note: [a] : input string; exp[b] : literature reported half-lives (days); pre[c]: half-life values calculated by the neural network.

Chemical and structure[a]	Half-life in days			
	High exp[b]	pre[c]	Low exp	pre
4-Aminoazobenzene 0 0 0 0 0 1 0 3 2 0 0 0 0 0	28	31.6	7	5.1
2 4-Aminobiphenyl 0 0 0 0 0 0 0 1 2 0 0 0 0 0	7	16.7	1	3.7
Amitrole 0 0 0 0 0 2 0 4 0 0 0 3 1 1	180	150.	28	16.8
2,4-Dichlorpheoxyacetic acid 1 0 1 1 0 0 2 0 1 0 0 0 0 0	50	19.4	10	5.7
Dieldrin 0 0 0 0 0 1 6 0 0 0 0 1 4 3	1080	1206.	175	68.8
alpha-Naphthylamine 0 0 0 0 0 0 0 1 2 1 0 0 0 1	180	316.	28	39.6
1,2,3-Trichloropropane 3 0 0 0 0 0 3 0 0 0 0 0 0 1	365	229.	180	47.2
1,2,4-Trimethylbenzene 3 0 0 0 0 0 0 0 1 0 0 0 0 0	28	22.3	7	3.8
Vinyl chloride 1 0 0 0 0 1 1 0 0 0 0 0 0 1	180	266.	28	53.4
Warfarin 1 2 1 1 1 1 0 0 2 0 0 1 1 0	28	17.4	7	2.0

138

6. References

1. Howard, P.H., Boethling, R.S., Stiteler, W.M., Meylan, W.M., Hueber, A.E., Beauman, J.A., and Larosche, M.E. (1992) Predictive model for aerobic biodegradability developed from a file of evaluated biodegradation data, *Environ. Toxicol. Chem.* **11**, 593-603.

2. Geating, J. (1981) *Literature study of the biodegradability of chemicals in water*, Vol. 1. Biodegradability, prediction, advances in and chemical interferences with wastewater treatment. EPA 600/2-81-175 (PB 82-100843). National Technical Information Service, Springfield, VA.

3. Boethling, R.S. (1986) Application of molecular topology to quantitate structure-biodegradability relationships. *Environ. Toxicol. Chem.* **5**, 797-806.

4. Desai, M.D., Govind, R., and Tabak, H.H. (1990) Development of quantitative structure-activity relationships for predicting biodegradation kinetics, *Environ. Toxicol. Chem.* **9**, 473-477.

5. Tabak, H.H. and Govind, R. (1993) Prediction of biodegradation kinetics using a nonlinear group contribution method, *Environ. Toxicol. Chem.* **12**, 251-260.

6. Desai, M.D., Govind, R., and Tabak, H.H. (1990) Development of quantitative structure-activity relationships for predicting biodegradation kinetics, *Environ. Toxicol. Chem.* **9** 473-477.

7. Gao, C., Govind, R. and Tabak, H. (1992) Application of the group-contribution method for predicting the toxicity of organic chemicals. *Environ. Toxicol. Chem.* 11: 631-636.

8. Gao, C., Govind, R., and Tabak, H. (1996) Predicting soil sorption coefficients of organic chemical using a neural network model. *Paper accepted for publication in Environ. Toxicology and Chem.*

9. Howard, P.H. (1985) Determining "real world" biodegradation rates, *Environ. Sci. Technol.*, **23**, 672-679.

10. Desai, S., Govind, R., and Tabak, H.H. (1990) Determination of Monod kinetics of toxic compounds by respirometry for structure-biodegradability relationships. *ACS Symp. Ser.* **422**: 142-156.

11. Govind, R., Gao, C., Yan, X., Pfanstiel, S. (1993) Development of a methodology for determination of bioavailability and biodegradation kinetics of toxic organic pollutant compounds in soil. *Paper presented at the In-situ and On-site Bioreclamation, 2nd International Symposium,* San Diego, CA, April 5-8.

12. Howard, H.P.; Boethling, S.R.; Jarvis, F.W.; Meylan, M.W.; Michalenko, M.E. (1991) *Handbook of Environmental Degradation Rates.* Lewis Publishers, Inc., Chelsea, MI, USA.

13. Lippmann, R.P. (1987) An introduction to computing with neural nets. *IEEE Trans. Acoust. Speech Signal Process.* **4**: 4-22.

14. Wasserman, P.D. (1989) *Neural Computing.* Van Nostrand Reinhold, New York, NY.

15. Bhagat, P. (1990) An introduction to neural nets. *Chem. Eng. Prog.* **86**: 55-60.

SUBJECT INDEX

Ab Initio calculations	61
Accessible surface area	94
AM1	75
Anaerobic conditions	19
Artificial intelligence	41
ASA	94
Back-propagation neural network	121
BCLU	78
BESS	65
Bioavailability	59
Biochemical transformation	4
BIODEG	41, 42, 111
Biodegradability testing	17, 18
Biodegradation pathways	65
Biodegradation rate constant	117
Biophore	31, 32
Bioremediation	115
Biosphere	7
Biphenyl	11, 55
BOD	20, 22, 93, 94, 95
CASE	30
CASETOX	27
Catechol pathway	29
Cell free extracts	77
Charge density	32
Classification & labelling	105
Cluster analysis	79
COD	23
CoMFA	94
Complete degradation	17
Correlation analysis	19
Cross-validation	109
Dead-end products	4
Dechlorination	51, 55,
Dechlorination pattern	55, 56
Dehalogenase	81
Dehalogenation	75
Dibenzo-p-Dioxins	51
Dibenzodioxin dioxygenase	13
Dioxins	51, 52
Dipole moment	78
Discriminant analysis	103, 108
Endogenous respiration	22

Environmental fate	105
Enzymatic kinetics	75
Enzymes	78
Expert judgement	41
Expert system	33, 66
External validation	111
Extracellular enzyme system	8
Fungal oxidation	9
Furan	54
Genetic algorithm	72
Half-life	116
Halidohydrolase	75
Haloaromatics	11
Hardness	78
Heat of formation	95, 97
Hexachlorobenzene	51
HOMO coefficients	32, 78
HOMO energy	78, 95
HOMO-LUMO energy gap	32
Hydrolase	81
Hydrolytic dehalogenation	84
Inductive machine learning method	41
Inter-species diversity	83
Intermediates	4
Intracellular enzyme system	8
Ionization potential	32, 78
Krebs cycle	11
Logical element	43, 44
LUMO energy	78, 95
Mechanistic approach	88
META	27
META-CASETOX	27
Metabolism	34
Microtox®	36
MITI	41, 42, 70, 93, 95, 105, 107
Model validation	105
Molecular connectivity	96
Molecular descriptors	76
Molecular fragment	76, 119
Molecular indicator	117
MOPAC	78, 93
MultiCASE	30
Multiple linear regression	94
Multivariate analysis	105
Neural networks	117
Nucleophilic delocalizability	78
Octanol/water partition coefficient (logP)	78

Oxygenase	81
P-450 enzymes	10
PAH	57
Partial degradation	17
Pattern recognition	59
PCA	52, 79
PCB	11, 55
PCDD	51, 52
PCDF	51, 52
Peroxidase	9, 10
PLS	62, 79, 108
Potential degradability	19
Primary degradation	17
Principal component analysis	52, 79
Priority substance	106
QSAR modelling	41
Quantum chemistry	75
Readily biodegradation	18
Risk assessment	1, 93, 105, 106
Risk evaluation	105
Rule generation system	41
SBR	3, 107
Sediment	51
Soil	51, 77, 115
Structural fragments	105
Substrate specificity	75, 83
Testing strategy	19
TOD	22, 23
Toxicity	4
Transformation kinetics	59
Transformation pathways	4
Ultimate biodegradation	41
Validation	109
Xenobiotics	7, 17

AUTHORS INDEX

P. Adriaens	51
A.L. Barkovskii	51
M. Cronin	93
J. Damborský	1, 51, 75
J. Dearden	93
L. Forney	65
D. Gamberger	41
R. Govind	115
B. Hansen	105
W. Karcher	105
G. Klopman	27
M. Kutý	51, 75
L. Lei	115
R. J. Larson	65
F. Lindgren	105
H. Loonen	105
M. Lynam	51
K. Manová	75
P.H. Masscheleyn	65
Ž. Medven	41
A. Patton	65
W.J.G.M. Peijnenburg	1
P. Pitter	17
B. Punch	65
A. Sabljić	41
S. Sekušak	41
V. Sýkora	17
H. H. Tabak	115
K. Wight	65
R.-M. Wittich	7

The manufacturer's authorised representative in the EU is Springer
Nature Customer Service Centre GmbH, Europaplatz 3, 69115 Heidelberg,
Germany. If you have any concerns regarding our products, please
contact ProductSafety@springernature.com

Printed and bound by CPI Group (UK) Ltd, Croydon, CR0 4YY

29/04/2026

02099460-0015